Steels: Metallurgy and Applications

Steels: Metallurgy and Applications

Second Edition

D. T. Llewellyn

Butterworth-Heinemann Ltd
Linacre House, Jordan Hill, Oxford OX2 8DP

A member of the Reed Elsevier plc group

OXFORD   LONDON   BOSTON
MUNICH   NEW DELHI   SINGAPORE   SYDNEY
TOKYO   TORONTO   WELLINGTON

First published 1992
Second edition 1994

© Butterworth-Heinemann Ltd 1992

British Library Cataloguing in Publication Data
Llewellyn, D. T.
    Steels: Metallurgy and applications – 2 Rev. ed
    I. Title   II. Series
    669.96142

ISBN 0 7506 2086 2

Library of Congress Cataloguing in Publication Data
Llewellyn, D. T.
    Steels : metallurgy and applications / D. T. Llewellyn
    p.   cm.
    Includes bibliographical references and index.
    ISBN 0 7506 2086 2
    1. Steel—Metallurgy.   2. Steel alloys—Metallurgy.   I. Title.
    TN730.L53   1992
    669'.142—dc20   91–37850
                                                             CIP

Every effort has been made to trace holders of
copyright material. However, if any omissions
have been made, the author will be pleased
to rectify them in future editions of the book.

Printed in Great Britain by Redwood Books, Trowbridge, Wiltshire

Contents

Preface

The first edition of this book was published in 1992 and I am delighted that it has sold well enough to merit a second edition in paperback which may be more affordable to the student population. However, the basic contents of the text are unchanged in covering the broad spectrum of the mainstream commercial steel grades and the manufacturing or service requirements that govern their application. Again the text assumes that the reader has a basic knowledge of ferrous metallurgy and the emphasis remains on alloy design and the generation of properties rather than on fundamental metallurgical concepts.

Since the first edition, major changes have taken place in the move to European steel specifications and certain British Standard specifications have been withdrawn. These changes have been recorded in this revision but, unfortunately, the process is incomplete and it has been necessary to refer to both British and European standards for some classes of steel grades. The data on steel prices have also been revised and now reflect the situation which was effective on 1 March 1994. Due to the depressed state of the market, most steel prices have increased only marginally from those recorded on 1 January 1991 and in the stainless steel sector, the prices have declined. However, it is again emphasised that such information can soon be out of date and prices have only been provided as a guide to the relative costs of grades in a given type of steel.

The author is indebted once more to many friends and former colleagues in the steel industry who provided information for this revised edition.

David T. Llewellyn
Department of Materials Engineering
University College of Swansea
May 1994

1 Technological trends in the steelmaking industry

Reconstruction of the industry

Following the oil crises of 1973 and 1979, there was a decline in the demand for steel and steelmakers throughout the world suffered major financial losses. Many will recall the situation in the UK, where in the financial year 1980/81, the British Steel Corporation (now British Steel plc) made a staggering loss of almost £700 million. However, throughout the 1980s, British Steel's financial performance improved dramatically and in 1989/90, it recorded a pre-tax profit of £733 million, making it by far the most profitable steelmaker in the world. This recovery, described by the *Sunday Times* as an *industrial resurrection*,[1] was all the more remarkable at a period of excess steelmaking capacity in Europe and the Western world, the problem being exacerbated further by the growth of steelmaking in the developing countries. However, the recession of the early 1990s and continuing over-capacity in Europe have reversed the financial trends of the 1980s, although figures published by British Steel for the first half of 1993/94 might indicate a return to profitability.

The reconstruction of the steelmaking industry in the UK is illustrated very forcibly by the manning and productivity data shown in Figure 1.1. At the time of the steel strike in 1980, British Steel employed more than 160,000 personnel but by 1990, the workforce had been reduced to less than 50,000. The social cost of this reconstruction has therefore been high with substantial works closures, many in areas where steel was the major employer. However, as illustrated in Figure 1.1, British Steel has achieved spectacular gains in productivity, making it one of the most cost-effective steelmakers in the world. In addition to demanning, technological changes have also played a very important part in the cost structure of the industry and some of these changes are outlined briefly in this chapter.

Energy conservation

Energy in the form of coal, oil, gas and electricity currently account for about 11% of the costs in steel production, being third in importance after raw materials and labour costs. Following the *oil shock* of 1979, energy prices increased spectacularly and therefore energy conservation became a vital part of British Steel's return to profitability. In 1980/81, energy consumption amounted to 29 GJ/tonne of liquid steel but by 1991/92, the figure had been reduced to 23 GJ/tonne. Whereas energy utilisation improved as loading was concentrated on fewer, more efficient plants, major savings were also achieved by carrying out detailed energy audits in specific process areas and by investment in energy-saving schemes that would provide a return in a reasonable time. Particular examples[2] of the schemes adopted during the early 1980s included:

1. The hot charging of continuously cast slabs directly to the reheating furnace

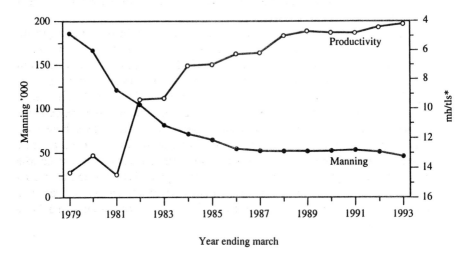

Figure 1.1 *Changes in manning and productivity – British Steel 1979–1993. Courtesy British Steel Technical*

whenever possible and the use of insulating covers when logistics dictate that the slabs must be stored.

2. The reduction of thermal losses during rolling by insulating the transfer table which connects the roughing and finishing stands at the Lackenby Coil Plate Mill. Normally, the significant temperature losses that occur during rolling must be compensated by a high *drop-out* temperature, but by avoiding the large energy loss on the transfer table, the average drop-out temperature was reduced by 55°C.

3. The installation of computer-based fuel economy systems in reheating furnaces which adjust the temperature difference between the furnace enclosure and the stock according to throughput. During steady furnace operation, the temperature difference is just sufficient to produce adequately

heated stock at discharge. However, at reduced throughput or during delays, the temperature profile is lowered to prevent overheating of the stock.

4. The use of *self-recuperative burners* and, more recently, *regenerative burners* for ladle heating and in reheating and heat treatment furnaces. In recuperative burners, combustion air is fed through an integral heat exchanger to recover energy from the waste gases that leave the furnace. The regenerative burner system operates in a pulsed manner and employs two burners with associated heat recovery equipment. Each burner is fired for a set period and waste gases are exhausted from the furnace by aspiration back through the burner which is not being fired. The system is capable of producing combustion efficiencies of over 70% whereas efficiencies can be as low as 30% in small conventional furnaces with poor or non-existent waste heat recuperation.[3]

5. The introduction of *expert systems* for more effective furnace management and also for other aspects of steel production. These are computer programs which operate on a database of the knowledge of human experience and offer major potential for improving product quality and reducing manpower. Expert systems have been applied to maintain the optimum air/fuel ratios in multi-burner furnaces, measurements of oxygen and carbon monoxide levels in the waste gases providing the basic parameters for a rule base.[3] As illustrated in Figure 1.2, nine sectors of inference are drawn from this rule base regarding burner performance and the actions necessary to correct the situation. Such systems use microprocessors to provide an automatic means of adjustment to individual burners via supplementary control valves. The savings from this technology are of the order of 2.5% for boilers and 5–20% for furnaces.

Ironmaking

The blast furnace continues to be the dominant facility for ironmaking, although during the 1960s and 1970s, major attention was given to the so-called *direct reduction* processes in which metallic iron is produced by the reduction of iron ore at temperatures below the melting point of the materials involved. Some of the processes were based on the use of inclined rotary kilns to which ore, coal and recycled materials are charged at the upper end and additional energy is supplied by some form of fuel burner. Other processes have involved gaseous reduction of which the Midrex process is the most successful. The operation of this process is illustrated in Figure 1.3. A gas reformer is employed to convert natural gas to reducing gas, containing about 95% ($CO + H_2$). In the shaft furnace, iron ore descends in counterflow to the reducing gas and is reduced to metallic iron at a temperature of about 850°C. The top gas containing about 30% ($CO_2 + H_2O$) is passed through a cooler-scrubber where steam and dust are removed and part of the gas is mixed with fresh natural gas before entering the reformer for the production of reducing gas. The direct reduction product contains about 95% iron together with unreduced oxides and sulphur levels that reflect the type of ore and fuels used in the process. Whereas

INFERENCE (3)	INFERENCE (6)	INFERENCE (9)
Most burners too lean	One or more burners too lean	One or more burners too lean One or more burners too rich
ACTION	ACTION	ACTION
Reduce air to all burners	Find and adjust lean burner (s)	As (8) followed by (6)

INFERENCE (2)	INFERENCE (5)	INFERENCE (8)
One or possibly more burners too lean	Operating at expectation	One or more burners too rich
ACTION	ACTION	ACTION
Find and adjust lean burner (s)	Reduce oxygen limits	Find and adjust rich burner (s)

INFERENCE (1)	INFERENCE (4)	INFERENCE (7)
Performance better than expectations	Performance slightly better than expectations	Most burners too rich
ACTION	ACTION	ACTION
Reduce oxygen limits slightly	Reduce oxygen limits slightly	Turn all burners slightly leaner

Figure 1.2 *The 'rule base' for fuel/air stoichiometry for furnace control by an expert system showing the nine actions that can be taken. Optimum operation is achieved when the measured O_2 and CO percentage values fall into Zone 5. If the values fall outside the limits defining this zone then control action must be taken. For example, high CO levels are shown together with low O_2 (Zone 7), the burner flame is too rich and the flow of combustion air must be increased by a calculated amount. (After Hibberd[2])*

direct reduction processes are an economic proposition for mini-mill plants, usually in developing countries, they have made little impact in established steelmaking countries.

More recently, attention has turned to an alternative ironmaking procedure known as *smelting reduction* of which the Corex process is an example. This has two main process units, namely a fluidized bed melter-gasifier and a counter-current reduction shaft. In the melter-gasifier, the injection of oxygen into a bed of coal char generates heat and also creates reducing gas. Directly reduced iron is fed into this unit where it melts, collects on the hearth and is tapped periodically. The gases from the melter-gasifier are strongly reducing and form the gas feed for the reduction unit where iron ore is converted to directly reduced iron.

During the 1980s, major attention was given to the practice of injecting auxiliary fuels, particularly coal, into blast furnace tuyeres. This leads to improved productivity but the overriding reason for coal injection in blast furnaces has been to reduce the costs of ironmaking, particularly when the coal has replaced purchased coke. However, other advantages[4] also accrue:

1. Steam is normally added to the air blast to moderate the temperature of the flame. The replacement of steam by coal injection decreases the rate of fuel

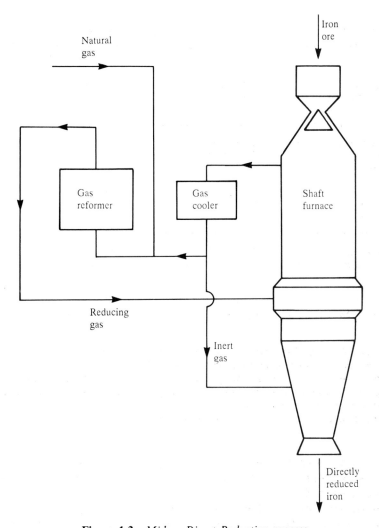

Figure 1.3 *Midrex Direct Reduction process*

consumption (*fuel rate*) and increases both the temperature and degree of oxygen enrichment of the blast.

2. Coal injection reduces the amount of coke to be added to the furnace and therefore increases the ore-charging capacity.

3. The energy consumption in the ironworks is reduced because less energy is required in preparing coal for injection than that required for producing an equivalent amount of coke.

Coal injection also reduces the demands imposed on aging coke ovens and, in some cases, alleviates the need for costly rebuilding programmes.

As described by O'Hanlon,[5] detailed attention has been given to improved process control in blast furnaces which has resulted in the productivity gains and reduction in fuel rate (equivalent coke rate) shown in figure 1.4. Several factors have contributed to the improved performance, including the

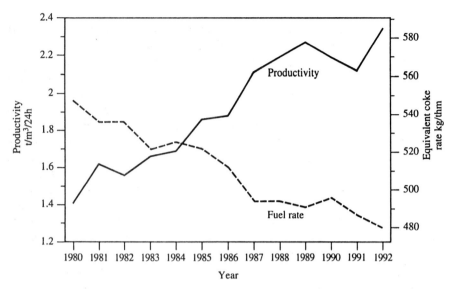

Figure 1.4   *Performance of British Steel's blast furnaces 1980–1992. Courtesy British Steel Technical*

rationalization of operating units and improvements in burden and coke quality. However, O'Hanlon identifies the major factor as the being the improvement in equipment, instrumentation and data-handling systems which enable operators to monitor the state of the process continuously and make the necessary corrections.

Steelmaking

Basic oxygen steelmaking

Basic oxygen steelmaking (BOS) now accounts for about 73% of steel production in the UK, the remainder being produced essentially via the electric arc furnace. Like other areas in the production route, steelmaking has undergone major developments which have been aimed at cost reduction and quality improvements in terms of steel cleanness and tighter compositional control.

In general, the main trend has been to reduce the refining burden on BOS furnaces by increasing the treatment carried out on hot metal before it enters the furnace and by completing the refining process in secondary steelmaking vessels. Thus the levels of silicon, manganese and phosphorus in blast furnace iron have been reduced and, in addition to shortening the steelmaking time, this has led to increased yields, reduced flux consumption and improved quality. The main desulphurization processes are also carried out on hot metal before it enters the BOS vessel with the addition of soda ash to the transfer ladle and the deep injection of calcium carbide or magnesium into the torpedo car. These practices, together with augmentation from secondary steelmaking, mean that linepipe specifications calling for 0.002% S max. can now be satisfied on a regular basis. However, lower levels are possible and Japanese steelmakers have declared their intention of producing steels with a total $(S+O+N+P)$ content of 0.0045% max.

One of the most important developments that has taken place in recent years in **BOS** technology has been the introduction of *bath agitation* processes. In these processes, the refining achieved with the top oxygen lance is supplemented by bubbling an inert gas, such as nitrogen or argon, through the base of the converter. This additional bath stirring accelerates the reactions that take place between the liquid metal and the slag, improves metallic yield and provides better process control. The implementation of BAP, the bath agitation process developed jointly by British Steel and Hoogovens BV, has been described by Normanton.[6] As illustrated in Figure 1.5, coal can be added to the vessel in order to increase the rate of scrap melting. Special lances have also been developed which supply a secondary source of oxygen above the level of the main jets, thereby increasing the combustion of CO to CO_2. It is claimed that scrap additions can be increased by 3 kg/tonne of liquid steel for every 1 kg of coal added with a further increase to about 5 kg/tonne if the secondary combustion lance is employed. Bath agitation can also improve steel quality by reducing the hydrogen content and by providing a facility for either reducing or increasing the nitrogen content.

A major feature of modern BOS practice is the high degree of automation and process control that is achieved through the use of computers. Process models ensure that the correct weights of hot metal, scrap, fluxes and oxygen are made to each cast to achieve the target composition and temperature. On-line process measurements then provide the means of controlling the blowing operation automatically. These include measurements of the composition and flow of the waste gases and the use of audiometers which monitor the state of the slag by

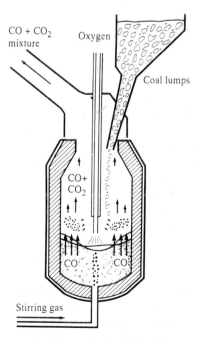

Figure 1.5   *The use of a secondary oxygen lance with in-blow coal additions in conjunction with BAP operation (After Normanton[6])*

measuring the damping effect of slag volume on the noise generated by the oxygen lance.[7] The oxygen lance is then moved automatically in order to correct errors in the carbon removal rate and slag volume.

Electric arc furnace steelmaking

The electric arc furnace (EAF) is used primarily for the production of carbon and alloy engineering steels and also as the primary melting stage in the production of stainless steels. The principal raw material for the EAF is recycled steel scrap which makes the process environmentally friendly and also energy efficient compared with production from iron ore. The latter requires additional energy to reduce the iron oxide to metallic iron and consumes about 20 GJ/tonne of steel compared with about 2.46 GJ/tonne for scrap remelting.[8] Modern EAFs have a capacity of 100–120 tonnes and operate at a power level of 70–90 megawatts.

Over the years, there has been a deterioration in the quality of steel scrap as the availability of the heavier grades of industrial scrap has declined and steelmakers have resorted to the use of lighter scrap from motor cars and consumer durables. This has increased the levels of *residual elements* such as copper and tin in the scrap which can only be controlled by dilution with purer and more expensive iron units. However, major effort and expenditure has been devoted to the development and installation of equipment that is capable of separating metallic from non-metallic scrap and segregating ferrous from non-ferrous scrap.

Like the BOS process, optimum conditions in the EAF are now achieved by computer control during the melting and refining stages. Control of melting is based on the measurement of the water temperature in water-cooled panels in the walls and roof. Oxygen input and oxy-fuel operations are also monitored and integrated with the electric power input to achieve optimum melting and economic performance.

Electrodes account for a significant part of the total cost of EAF operations, a figure of £7.5/tonne of liquid steel being quoted in 1985.[9] However, the development of water-cooled electrodes has reduced the components of wear due to oxidation and mechanical/thermal shock. The use of water-cooled panels in the furnace has also contributed to longer life and reduced electrode costs by permitting the use of longer arcs without the attending problems of excessive refractory wear. Longer arcs result in higher voltages and lower currents for a given power input and the reduced current densities result in lower tip losses.

Oxy-fuel burners are now employed on EAFs to provide auxiliary heating and this has resulted in improved thermal efficiency. These burners, fired with oil, gas or coal, are normally directed at the relatively cold spots in the furnace, i.e. between the areas that are more directly heated by the electrodes. The combined use of energy from the arcs and oxy-fuel burners is most efficient in the initial stages of melting when the furnace is full of scrap. Electricity is the major item of cost in EAF operation and it is anticipated that further developments will take place to increase the substitution of electricity by fossil fuels, particularly coal.

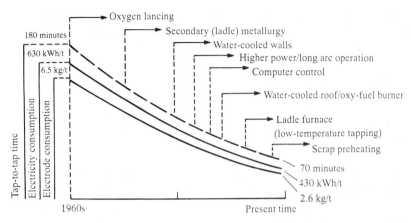

Figure 1.6 *Some EAF developments and how they have helped to improve energy and electrode consumption and reduce melting time (After Smutts-Müller et al.[8])*

The various developments that have taken place since 1960 and which have contributed significantly to the increased productivity and energy efficiency of EAFs are summarized in Figure 1.6.[8]

Secondary steelmaking

In recent years, there has been a marked increase in the use and variety of secondary steelmaking practices for the following reasons:

1. The demand for higher quality steel products with lower levels of elements such as hydrogen, carbon, nitrogen, sulphur and phosphorus.
2. The need for improved cleanness and lower levels of segregation.
3. The requirement for closer control over composition and temperature, particularly in relation to subsequent continuous-casting operations.

Better control over composition and temperature is achieved by stirring the steel in the ladle with argon for two or three minutes. Argon is normally introduced via a lance which is inserted deep into the ladle although porous plugs in the ladle bottom have also been used. Following the analysis of a representative sample, the composition is then adjusted (*trimmed*) by the addition of ferro-alloys and the temperature of the steel is controlled by further stirring or by adding steel scrap.

In order to achieve very low levels of sulphur, i.e. less than 0.002%, mixtures of lime and fluorspar are added to the steel during tapping to produce a synthetic basic slag on the steel in the ladle. Synthetic lime-based powders are also injected into the steel in the ladle. Where sulphur levels of less than 0.001% are required, calcium silicide is injected into the steel, either as a powder or in the form of wire. As well as reducing the sulphur level, calcium additions are also beneficial in modifying the inclusion composition or morphology, i.e. by converting alumina to calcium aluminate and by forming more globular manganese sulphides.

Although vacuum degassing units were introduced initially to lower the hydrogen content of the liquid steel, they have also proved to be very effective in controlling the carbon and oxygen contents and are used extensively as stirring and trimming stations.

Ladle reheating is one of the more recent innovations in the field of secondary steelmaking and is proving to be particularly effective in achieving improved compositional and temperature control. Reheating is achieved by means of electrodes which are inserted through the roof above the ladle, such that the ladle is transformed to a small electric arc furnace. The facility for reheating in the ladle permits a reduction in the furnace tapping temperature which allows lower phosphorus levels to be achieved during steelmaking and also reduces refractory costs. Ladle reheating also affords the opportunity for making larger alloy additions to the ladle, thereby increasing the range of steels that can be produced via the BOS route. Benefits accrue too from the fact that the steel can be held for longer times in the ladle, thereby relieving the transfer problems between the steelmaking furnace and casting machine.

Continuous casting

Production data for 1989 indicated that 82% of the steel output in the UK was continuously cast and it is predicted that the level will soon exceed 90%. However, it is unlikely that ingot casting will be eliminated completely since there will continue to be a demand for ingot production in special products such as large rotor forgings and heavy plates.

In addition to the very significant gains in yield and energy savings compared with the ingot route, continuous casting has also resulted in major improvements in quality. However, the maintenance of high quality standards in continuously cast products requires very careful attention to detail at all stages in the process. For example, the exposure of the liquid steel to the atmosphere can undo much of the refining achieved in secondary steelmaking by allowing the steel to pick up oxygen and nitrogen. For this reason, shrouds are employed to protect the metal stream as it is poured from the ladle to the tundish and from the tundish to the mould. In many cases, shrouding is provided by a curtain of inert gas such as argon or carbon dioxide but refractory tubes are also used for this purpose.

Great care is also exercised to avoid the entrainment of non-metallic inclusions at various stages in the casting process. This involves the use of detectors which prevent slag carryover from the ladle to the tundish and special devices which control the levels of liquid in both the tundish and mould. Improvements in submerged entry nozzle design have also resulted in a marked reduction in rejections due to entrained mould slag.[10]

Product requirements

From the foregoing remarks, it can be appreciated that steelmaking has now become a hi-tech industry and that the adoption of modern technology has resulted in major gains in productivity and product quality. This has played a vital part in the recovery of the steel industry in the UK and has helped to improve the cost-effectiveness of steel relative to competitive materials such as plastics and aluminium.

Each of the following chapters starts with a brief *Overview* in which an attempt has been made to summarize some of the important changes that have occurred in the product requirements for various types of steel. Improvements in steel cleanness and compositional control feature prominently in most products but there has also been a major call for steels that will reduce the cost of the final component. In some cases, this involves the use of higher strength steels which afford the opportunity for down-gauging, whereas in others the steels supplied result in a decrease in heat treatment or fabrication costs. However, steelmakers cannot afford to be complacent and further developments will be required if steel is to remain a dominant engineering material well into the 21st century.

References

1. Lorenz, A. *Sunday Times*, 14 October 1990.
2. *Steelresearch 84–85*, British Steel, 1 (1985).
3. Hibberd, D.F. *Steelresearch 87–88*, British Steel, 27 (1988).
4. *Steelresearch 84–85*, British Steel, 8 (1985).
5. O'Hanlon, J. *Steelresearch 87–88*, British Steel, 7 (1988).
6. Normanton, A.S. *Steelresearch 87–88*, British Steel, 12 (1988).
7. Barradell, D.V. *Steelresearch 87–83*, British Steel, 17 (1988).
8. Smutts-Müller, D., Nicholson, A. and Wilson, E. *Chemistry Now – Electric Arc Furnace Steelmaking*, Hobsons Publishing, Cambridge (1990).
9. *Steelresearch 84–85*, British Steel, 11 (1985).
10. *Steelresearch 82–83*, British Steel, 12 (1983).

2   *Low-carbon strip steels*

Overview

The first continuous hot strip mill was commissioned in 1923, in the United States, and revolutionized both the steel industry and the market for strip products. This development made available wide steel strip at a lower price and with properties far superior to those offered previously by the old, hand-operated mills and, in particular, resulted in the dramatic growth of the automotive industry. In turn, this industry has called for ever-increasing standards of quality in terms of gauge control, flatness, surface texture and cleanness and has been the stimulus for major product development in the strip area.

These steels are produced via large BOS vessels in integrated steelworks and are available in both the hot-rolled and cold-reduced conditions. At the present time, hot-rolled material can be produced in thicknesses down to about 2 mm but the main demand is for material which has been cold reduced and softened in batch or continuous annealing furnaces. The main property requirement of these steels is a high level of cold formability and therefore a great deal of strip is produced with carbon contents of less than 0.05% and with manganese contents of the order of 0.20%. However, good formability also requires the generation of particular crystallographic textures which are induced by control of composition and processing conditions.

In recent years, the automotive industry has called for the development of higher strength steels that will allow the down-gauging of body panels so as to reduce vehicle weight and improve fuel economy. This has presented a major challenge in terms of providing higher strength steels with adequate levels of formability and also in dealing with the problem of *springback* (elastic recovery). The automotive industry has also been forced to improve its warranties against corrosion in vehicles and this has led to a dramatic increase in the use of zinc-coated steels such that these materials now account for about 70% of the strip requirements of most motor cars.

Developments in the building industry have also had a major impact on the demand for wide strip products. In particular, there has been a spectacular growth in the use of profiled, organic-coated, galvanized sheet for architectural roofing and cladding and the galvanized steel lintel has largely replaced reinforced concrete lintels in domestic housing. On the other hand, alternative materials have penetrated the traditional steel-based applications, a notable example being the use of plastics in automotive bumpers. However, it is the view of one car maker in the UK that the steel industry has responded well to the challenges posed by aluminium and polymers and that, unless a major oil or energy crisis arises, steel will remain the first-choice material for motor car panels for many years to come.[1]

Cold-forming behaviour

Cold formability represents the single most important property requirement in low-carbon strip grades and various laboratory tests have been devised in order to evaluate their cold-forming behaviour. These tests can examine the performance of steels under specific modes of deformation whereas others are designed to be simulative of the more complex straining behaviour in commercial pressings. These tests are reviewed and a brief insight given into the developing technique of applying finite element modelling to cold-forming behaviour.

True stress–true strain

Whereas there is general appreciation of *engineering stress–strain* curves, the concepts of *true stress* and *true strain* are not so well known and need to be introduced prior to a description of laboratory tests for sheet metal forming.

In conventional or engineering stress–strain curves, the stress is defined as the instantaneous load divided by the original cross-sectional area of the test piece. In these curves, the stress is shown to increase up to the tensile strength, i.e. the point of maximum load, but then decreases as the specimen undergoes local necking and is no longer capable of sustaining such high loads. On the other hand, true stress is defined as the instantaneous load divided by the instantaneous cross-sectional area of the test piece. Thus, as illustrated in Figure 2.1, the true stress continues to increase with increasing strain and represents the actual stress on the specimen right up to the point of fracture.

If a cylinder of length L_o is strained in tension to twice its original length, then by convention the engineering strain is defined as:

$$\frac{2L_o - L_o}{L_o} = 1.0 \ (100\%)$$

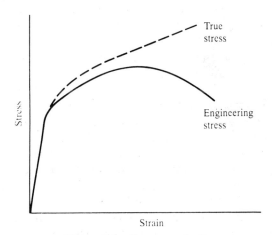

Figure 2.1 *Stress–strain diagrams*

If this procedure is then reversed, i.e. the elongated cylinder of length $2L_o$ is compressed to length L_o, then logic would indicate that the specimen should be strained in compression by 100%. Very clearly, this type of approach does not provide a satisfactory answer for situations involving both tension and compression, and to overcome the problem the concept of true strain was introduced. This defines strain as the change in length compared with instantaneous length and the total true strain as the following summation:

$$\varepsilon = \Sigma \; \frac{L_1 - L_o}{L_o} + \frac{L_2 - L_1}{L_1} + \frac{L_3 - L_2}{L_2}$$

$$= \int_{L_o}^{L} \frac{dL}{L}$$

$$= \ln \frac{L}{L_o}$$

Thus when a cylinder of length L_o is strained in tension to length $2L_o$:

$$\varepsilon = \ln \frac{2L_o}{L_o} = Ln\,2$$

Conversely, when a cylinder of length $2L_o$ is compressed to length L_o:

$$\varepsilon = \ln \frac{L_o}{2L_o} = \ln \tfrac{1}{2} = -\ln 2$$

Therefore true strain provides the same numerical value in both tension and compression but of opposite sign.

Deep drawing

Deep drawing is illustrated schematically in Figure 2.2(a) and involves the uniform flow of material into a die under a hold-down pressure that is just sufficient to prevent wrinkling. This requires that the material shall flow easily in

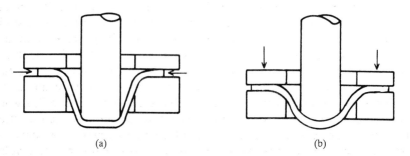

(a)   (b)

Figure 2.2   *(a) Deep drawing; (b) stretch forming*

all directions in the plane of the sheet and resist local thinning in the side walls during elongation. The anisotropy of a material has a marked influence on the distribution of strain that can be obtained in a pressing and Lankford *et al.*[2] in the 1940s developed the concept of the strain ratio (*r value*) to express plastic anisotropy. The *r* value is the ratio of the true strain in the width direction to the true strain in the thickness direction in a tensile test. Thus:

$$r = \frac{\varepsilon_w}{\varepsilon_t}$$

The strain ratio is related to crystallographic texture in sheet steels and varies with test direction in anisotropic materials. Therefore *r* values are measured in directions parallel, transverse and at 45° to the rolling direction and the $\bar{r}$ value represents the average plastic anisotropy of the material:

$$\bar{r} = \frac{r_1 + 2r_{45} + r_t}{4}$$

It has been shown that $\bar{r}$ (also designated r_m) is related directly to the maximum size of a blank that can be drawn to form a flat-bottomed cup.

Hot-rolled strip is fairly isotropic with an $\bar{r}$ of about 1.0 whereas aluminium-killed cold-reduced steel has a value of about 1.6. Deeper draws and more complex shapes can be achieved with a well developed crystallographic texture as indicated by high $\bar{r}$ values.

Differences in planar isotropy are generally expressed as:

$$\Delta r = \frac{r_1 + r_t - 2r_{45}}{2}$$

and this expression provides an indication of the amount of peaking or *earing* that will occur on the edges of deep-drawn cups. An isotropic material has a Δr value of zero and materials with values approaching zero show little tendency for earing.

Stretch forming

The deformation mode in stretch forming is illustrated in Figure 2.2(b). In this case, the hold-down pressure on the flange is high enough to prevent material flowing into the die cavity and the sheet deforms by elongation and uniform thinning as the punch descends. The material must therefore be capable of withstanding large uniform elongation before the onset of necking or plastic instability. The parameter which is accepted widely as an indicator of the stretch-forming behaviour is the *strain-hardening exponent* (also called work-hardening coefficient) or *n value* which represents the slope of the true stress versus true strain curve in a tensile test, plotted on logarithmic coordinates. As shown in Figure 2.3, the relationship can conform to a reasonably straight line and the data expressed by the following equation:

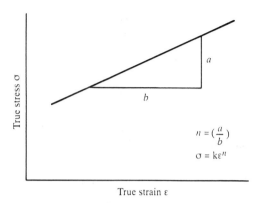

Figure 2.3   *Log–log plot of true stress versus true strain*

$$\sigma = k\varepsilon^n$$

where σ is the true stress, ε the true strain and k and n are constants. It can be shown that n is numerically equal to the uniform elongation, i.e. the elongation that occurs before the onset of necking. High values of n indicate good stretch formability and cold-reduced formable grades have n values in the range 0.22–0.25. On the other hand, hot-rolled materials can have n values as low as 0.1 and may undergo excessive thinning or fracture in severely strained regions of a pressing. Materials with high n values can work harden sufficiently in critical areas to transfer the strain to adjacent areas, thereby avoiding high local concentrations of strain.

Bending

This mode of deformation normally requires simpler tooling and can often entail less severe strains than those involved in deep drawing or stretch forming. The bending characteristics of sheet materials are generally measured by forming around rods of successively smaller diameter until cracking is produced at the *minimum bend radius*. However, it is usual to express this property as multiples of the sheet thickness. Thus a ductile material may be capable of producing 1*t* bends, where a less formable material might only be capable of achieving 3*t* bends.

Forming limit diagrams

Whereas values of $\bar{r}$, n and uniform elongation give good indications of forming behaviour under pure drawing or stretching, commercial pressings invariably involve mixed and more complex modes of deformation. As a means of illustrating the behaviour in commercial pressings, Backhofen and Keeler[3] introduced the concept of *forming limit diagrams* (FLDs). The layout of an FLD is shown in Figure 2.4.

A grid pattern, normally an array of circles of about 2 mm diameter, is first etched onto the surface of a sheet by means of a stencil and electrochemical or

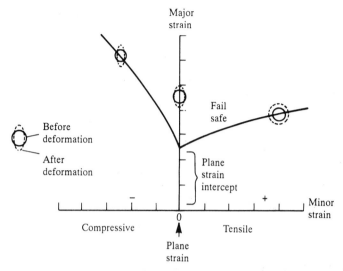

Figure 2.4 *Forming limit diagram*

photographic marking techniques. On straining, the circles become elliptical and the major strain, ε_1, is defined as the strain in the plane of the sheet along the direction of maximum strain. The minor strain, ε_2, is the strain in the plane of the sheet perpendicular to the direction of maximum strain. The major strain is always regarded as a positive variable and plotted on the vertical axis of the FLD. The minor strain is plotted horizontally and can be either positive or negative, depending upon whether the material has undergone an expansion or contraction in that direction.

A variety of forming operations may be used to define an FLD for a given material, the major and minor strains being measured at locations immediately adjacent to failure. The locus of critical strain combinations represents the forming limit and provides the type of curve shown in Figure 2.4. The area under the curve gives the range of strain combinations that the material can safely withstand. Newly designed pressings are often produced initially from grid-marked sheets and the strains compared with those shown to be critical in the FLD. This might indicate that a particular design of pressing represented a near-failure condition in a particular material, necessitating a modification in design or a more formable grade. As illustrated in Figure 2.5, the intercept of the FLD on the major strain axis is a function of sheet thickness and the strain-hardening exponent.

Finite element modelling

Although the FLD provides a valuable guide to forming behaviour and can assist in component design, the approach is still empirical. Additionally, it has limitations by virtue of the fact that the strain paths in commercial pressings may be significantly different from those encountered in the simple laboratory tests that are used to construct the FLD. Therefore there has been a move to

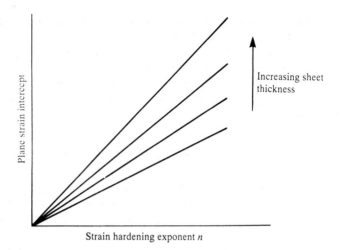

Figure 2.5   *Relationships between plane strain intercept and strain hardening exponent as a function of sheet thickness*

eliminate the empiricism involved in the approaches described above through the use of finite element modelling. It is now possible to simulate a wide variety of different forming processes by computer-aided engineering and Miles[4] has described the use of this technique in drawing operations on low-carbon steel. This is illustrated in Figure 2.6, which shows the computer simulation of a deep-drawn cup. The purpose of this particular experiment was to predict the tool forces required to form the component, the hold-down pressure on the blank and the tearing tendency of the sheet if the friction at the blank holder stopped the material from drawing correctly. Using this approach, it is also possible to

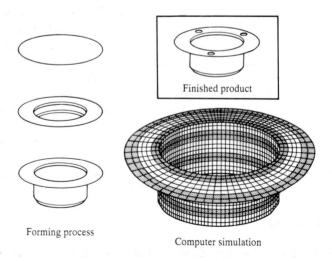

Figure 2.6   *Computer simulation of deep-drawing operations (After Miles[4])*

identify areas of thinning and thickening in the sheet and where excessive compression might lead to wrinkling.

The computer programs that have now been generated for this type of work are extremely powerful and versatile and permit the analysis of a wide variety of three-dimensional shapes. The effects of planar and normal isotropy can also be accommodated, together with parameters such as friction coefficient and strain rate sensitivity.

Metallurgical factors affecting cold formability

Major research work has been carried out on the metallurgy of formable, low-carbon strip and the literature on this topic has been reviewed by Hudd.[5,6] In general, the published work relates more to the factors affecting deep drawing and $\bar{r}$ values and there appears to be relatively little information on the metallurgical variables controlling stretch forming and n values. Because the method of annealing places very different demands on the prior processing and compositional control required for good formability, the topic is best discussed under the separate headings of batch annealing and continuous annealing. However, the understanding of these requirements will be facilitated by brief reference to the production aspects of low-carbon strip.

Process route

A typical process route for low-carbon strip is shown in Figure 2.7. Following secondary steelmaking, the liquid steel may be ingot cast but the bulk of current production is continuously cast. This being the case, the steel will be aluminium killed and, in turn, this has a very significant effect on the processing requirements for good formability. At typical slab-soaking temperatures around 1200–1250°C, aluminium nitride is in solid solution and will normally remain in this state after the completion of hot rolling. The hot-rolled material will then be subjected to a coiling operation which represents a very critical stage in the process route. If coiling is performed at a relatively high temperature, i.e. around 710°C, the large, tightly wound coil with a high thermal mass will cool very slowly and afford the opportunity for the precipitation of aluminium nitride. On the other hand, if the hot-rolled material is cooled quickly to a temperature of about 560°C before coiling, the precipitation of aluminium nitride is suppressed and both aluminium and nitrogen will be retained in solid solution on cooling to room temperature.

Hot rolling can be undertaken down to a thickness of about 2 mm but the production of thinner material requires further deformation by cold rolling. Following cold rolling, the material will be subjected to annealing heat treatments or routed for the production of tinplate or galvanized strip.

Batch-annealed strip

It is established that the deep-drawing characteristics of low-carbon strip are influenced significantly by the crystallographic texture. Good drawability is

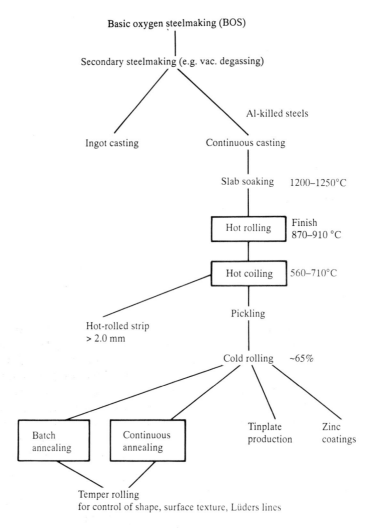

Figure 2.7 *Process route – low-carbon steel strip*

obtained by the development of a strong {111} cube on corner texture and the reduction of the {100} cube on face component. Rimming steels normally give $\bar{r}$ values in the range 1.0–1.2 but values of up to 1.8 can be produced in aluminium-killed steels. The addition of aluminium is beneficial to formability due to the generation of a favourable texture and a large ferrite grain size. However, in order to be effective in promoting these features, the aluminium must be present in the steel in solid solution prior to the annealing operation. Therefore, material destined for subsequent batch annealing will be coiled at a low temperature, typically 560°C, in order to avoid the precipitation of aluminium nitride.

The heat treatment cycle in batch annealing is shown schematically in Figure 2.8. Because of their large mass, the heating and cooling rates in the coils are very slow and the batch-annealing operation might proceed over several days,

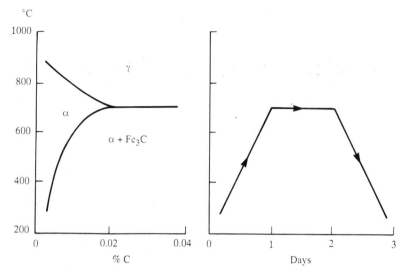

Figure 2.8   *Batch annealing*

depending upon the level of formability that is required. The strip is heated slowly to a temperature of about 700°C, close to Ac_1, and recrystallization of the cold-worked structure will take place in the temperature range 500–550°C. However, during the initial heating process, aluminium nitride precipitates on the deformation sub-grain boundaries which retards the recrystallization process, inhibiting the nucleation of new grains and thereby producing a large grain size. At the same time, the precipitation of aluminium nitride also induces the formation of a strong {111} texture and the effect is clearly dependent upon the heating rate and also on the relative proportions of aluminium and nitrogen. It has been shown[7] that the highest $\bar{r}$ values are produced in steels containing 0.025–0.04% Al and 0.005–0.01% N.

During the soaking period, grain growth proceeds and batch-annealed material develops a relatively coarse grain size of ASTM 5–6. Although higher soaking temperatures would be beneficial in this respect, they are limited to a maximum of about 730°C in order to avoid the formation of coarse carbides, which are deleterious to formability. In addition, higher soaking temperatures also increase the risk of the sticking of adjacent laps in a coil. As indicated in Figure 2.8, the ferrite microstructure will contain the maximum amount of carbon in solid solution at a soaking temperature of 700°C but the slow rate of cooling from this temperature results in the precipitation of most of the carbon on cooling to ambient temperature. Therefore, batch-annealed, aluminium-killed steel is characterized by:

1. A strong {111} texture.
2. A large ferrite grain size.
3. Low solute carbon and nitrogen contents.

However, the cooling rate can be adjusted to retain some carbon in solid solution which offers the potential for a strengthening reaction during subsequent paint-stoving operations. Such a process is termed bake hardening and will be discussed later in the text in a section dealing with higher strength steels.

Continuous annealing

The first application of continuous annealing for steel strip was introduced by Armco Steel Corporation in the United States for hot dip galvanized steel in 1936. Since that time, various facilities have been introduced for the continuous annealing of aluminized steel, tinplate, stainless steels and non-orientated silicon steel. Continuous annealing offers several production and quality advantages over batch annealing, including more uniform properties, cleaner surfaces and shorter production times. In spite of these advantages, continuously annealed strip was not considered suitable for automotive applications for many years because the cold-forming properties and resistance to aging were inferior to those achieved in batch-annealed material. However, in the early 1970s, Japanese steelmakers incorporated an overaging treatment in the continuous-annealing process and demonstrated that very much improved properties could be produced.

As described earlier, the batch annealing of aluminium-killed steel involves the precipitation of aluminium nitride during the slow-heating phase of the process. Due to kinetic considerations, this is not possible under the fast heating rates employed in continuous annealing and nitrogen remaining in solid solution would lead to increased strength, reduced formability and susceptibility to strain aging. In order to reduce the level of nitrogen in solid solution, hot band material which is scheduled subsequently for continuous annealing is coiled at high temperatures, i.e. at temperatures up to 710°C, depending upon the grade. Thus the material is able to cool slowly in coil form through the temperature range that favours the precipitation of aluminium nitride and the removal of nitrogen from solid solution.

The heat treatment cycle for continuous annealing is shown schematically in Figure 2.9. The process is characterized by rapid heating, a short soaking time and rapid cooling such that the process can be completed in four to eight minutes. Because of the short duration, there is little danger of sticking in the coil or the opportunity for carbide coarsening and therefore the maximum annealing temperature can be raised above Ac_1 in order to encourage grain growth. The rapid rate of cooling gives little time for carbide precipitation and therefore an overaging stage is introduced into the cycle in order to reduce the carbon content to an acceptably low level. Typical stages in the cycle include heating to 700–850°C in less than one minute, holding for 40 seconds, cooling to an overaging temperature range of 400–450°C and holding for up to three minutes, and finally cooling to ambient temperature. In spite of the overaging treatment, some carbon remains in solid solution and continuously annealed strip can display aging effects. In order to minimize this effect, it is beneficial to reduce the carbon content from normal levels of about 0.04–0.05% in batch-annealed material to levels around 0.02–0.03% in continuously annealed strip.

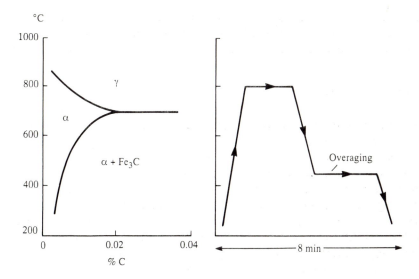

Figure 2.9 *Continuous annealing*

Compositional effects

Ono *et al.*[8] have investigated the effect of carbon on the mechanical properties of continuously annealed material and an illustration of their work is shown in Figure 2.10. This shows that the *n* value increases progressively with a reduction in carbon content whereas the $\bar{r}$ value appears to reach a plateau as the carbon content is reduced to about 0.02%. On the other hand, reduction in carbon below this level causes an increase in the aging index, and yield and tensile strength, and a reduction in the elongation values. These effects are related to the low driving force for carbide precipitation during overaging in compositions containing a low solute carbon content. This effect is illustrated in Figure 2.11, which shows the relationship between solute carbon content before and after overaging. In materials which are air cooled or water quenched after overaging, the amount of carbon in solution after overaging shows a maximum at an initial carbon content of about 60 ppm. Thus overaging will result in little precipitation of carbide in materials containing less than 60 ppm carbon in solution before the overaging treatment, and because of these effects it is generally undesirable to reduce the carbon content to below 0.02% during steelmaking.

Because of the short cycle times, continuously annealed materials tend to have a small grain size which in turn adversely affects the $\bar{r}$ values. Although these effects are compensated to some degree by raising the annealing temperature, the removal of elements from solid solution has also been shown to be beneficial in promoting grain growth. It is therefore considered desirable to restrict the manganese content to a minimum level which is consistent with adequate deoxidation and fixation of sulphur.[9] This effect is illustrated in Figure 2.12, which shows the relationships between $\bar{r}$ and the *K* value which is defined as follows:

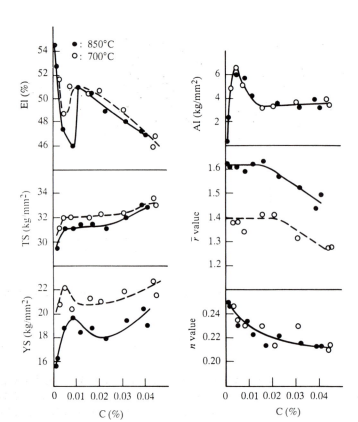

Figure 2.10   *Effect of carbon content on the mechanical properties; YS = yield strength; TS = tensile strength; AI = aging index (After Ono et al.[8])*

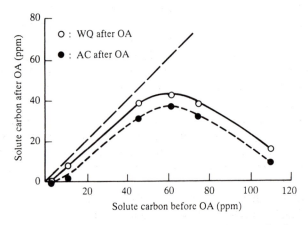

Figure 2.11   *Relation between solute carbon content before and after overaging; WQ = water quenching; AC = air cooled; OA = overaging (After Ono et al.[8])*

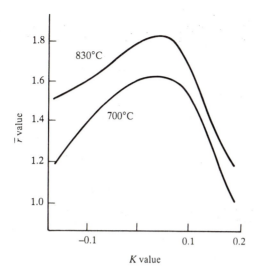

Figure 2.12 *Effect of* k *value and annealing temperatures on* r̄ *value (After Toda et al.[9])*

$$K = Mn\% - \frac{55}{32}S\% - \frac{55}{16}O\%$$

When $K=0$, manganese is at the stoichiometric level for combination with sulphur and oxygen and this provides an r̄ value near the optimum. When K is negative, some sulphur and oxygen are in solid solution, whereas when K is positive, excess manganese is in solid solution.

As indicated earlier, nitrogen in solid solution is detrimental to cold formability and continuously annealed material is hot-coiled at a high temperature in order to induce the precipitation of aluminium nitride. However, hot coiling has disadvantages in terms of producing an oxide which is difficult to remove during pickling and also because the non-uniformity in cooling rates in the strip tends to increase as the coiling temperature increases. Thus coil ends and edges cool faster than the body of the coil, resulting in finer grains in these regions. However, it has been proposed that the use of high-temperature coiling can be avoided through the addition of small amounts of boron.[10] Boron has a stronger affinity for nitrogen than aluminium, and by adding a suitable amount of boron (B:N = 0.8–1.0), most of the nitrogen is fixed as boron nitride, irrespective of the coiling temperature.

In a similar vein, additions of niobium and titanium can be used to reduce the carbon and nitrogen solute contents to very low concentrations and thereby improve the cold-forming properties of low-carbon strip. Such steels are often referred to as *interstitial-free (IF)* and a typical composition is given below:

C%	Mn%	Al%	Ti%	N%
<0.003	0.18	0.04	0.04	0.002

In such steels, the presence of titanium carbides raises the recrystallization temperature, and for this reason the steels are finished rolled above 950°C. IF steels are characterized by high $\bar{r}$ values, e.g. about 2.0, and provide an option for achieving good forming properties in continuously annealed steels.

High-strength strip steels

There is major interest in weldable, formable, high-strength strip, in both the hot-rolled and cold-reduced conditions, and a large number of grades has been developed by individual steelmakers throughout the world. According to an IISI publication,[11] a cold-reduced steel would normally be regarded as having high strength if the yield strength exceeded ~ 220 N/mm^2 or if the tensile strength exceeded ~ 330 N/mm^2. For a hot-rolled steel, the equivalent values are ~ 280 and 370 N/mm^2 respectively. The main driving force for the introduction of these steels has been the ability to use lighter gauges than those used tradition- ally in carbon steel grades, whilst still maintaining the strength and integrity of the component. As illustrated later, higher strength steels are now being used extensively in the automotive industry where weight saving has become particularly important but there is also a significant market for high-strength, galvanized steel in the construction industry.

Various metallurgical options are employed for the development of higher strengths in strip grades, depending upon the level of strength required and whether the strip is produced in the hot-rolled or cold-reduced condition.

Precipitation-strengthened/grain-refined steels

As illustrated in Chapter 3, a considerable amount of work has been carried out on the strengthening of ferrite–pearlite steels through the joint action of ferrite grain refinement and precipitation strengthening. The effect is achieved through the addition of small amounts of carbide- or carbonitride-forming elements such as niobium, vanadium and titanium (*micro-alloy additions*), and in both the structural and strip grades, the materials are known as *High-strength Low- alloy* (HSLA) steels.

At slab-soaking temperatures around 1200°C, the addition of micro-alloying elements restricts the size of the austenite grains through the presence of undissolved particles such as Nb(CN) or TiC. However, a proportion of these elements is taken into solid solution at the soaking temperature and this has two important effects. Firstly, these elements inhibit recrystallization during hot rolling which produces a fine austenite grain size and which in turn induces a fine ferrite grain size. Secondly, fine carbide/carbonitride particles are precipi- tated at the austenite–ferrite interface (*interphase precipitation*) on cooling to ambient temperature. Thus hot-rolled materials can be strengthened by the separate mechanisms of grain refinement and precipitation strengthening. The magnitude of the effects is dependent on the type and amount of element added and also on other aspects such as base composition, soaking temperature, finishing temperature, coiling temperature and cooling rate to ambient tempera- ture. However, as indicated in Figure 2.13,[12] strengthening increments of up to

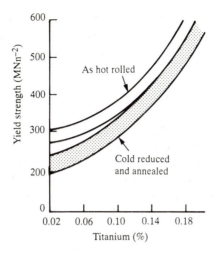

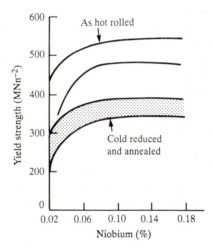

Figure 2.13 *Effect of niobium and titanium additions on yield strength of hot strip and cold-reduced strip after temper rolling (After Bordignon et al.[12])*

300 N/mm² may be achieved such that yield strength values of 500–600 N/mm² can be produced in the hot-rolled state.

The fine precipitates that are developed in hot-rolled strip are also instrumental in maintaining a higher strength in material that is subsequently cold rolled and annealed. However, they act mainly to refine the ferrite grain size during the recrystallization of the cold-worked structure and the coarsening of the particles during the annealing operation eliminates the precipitation-strengthening effect. Therefore the strength levels developed in cold-reduced strip are lower than those of parent hot band material. As illustrated in Figure 2.12, yield strength values of the order of 350 N/mm² are produced in cold-reduced strip, containing 0.06–0.1% Nb.

Solid solution strengthened steels

Elements such as manganese, silicon and phosphorus exert significant strengthening effects in ferrite but, in the context of low-carbon strip grades, strengthening with phosphorus has received the greatest attention. Typically, additions of up to 0.1% P are incorporated in *rephosphorized* auto-body strip with little extra cost or sacrifice in formability. The strengthening effect is of the order of 10 N/mm^2 per 0.01% P and the material is generally supplied to the automobile industry with yield strengths in the range 220–260 N/mm^2 and with $\bar{r}$ values of the order of 1.6.[13] The strength levels are obviously modest compared with HSLA grades but the phosphorus content is restricted to 0.1% max. as larger additions would impair the weldability of the material.

As indicated earlier in the section dealing with cold formability, interstitial-free (IF) steels have been developed for parts involving severe deep-drawing operations. The effective carbon and nitrogen contents of these steels are reduced to very low levels by means of stabilizing additions of titanium or niobium. However, these steels are relatively weak and are sometimes strengthened with small additions of phosphorus, manganese and silicon. Such steels, designated IF-HSS, maintain $\bar{r}$ values around 2.0 and provide tensile properties similar to those obtained in aluminium-killed, rephosphorized grades.

Dual-phase steels

In the late 1970s, major interest was generated in the United States in low-alloy steels that were heat treated specially to form a mixed microstructure of ferrite and martensite. The materials were designated *dual-phase* steels and are characterized by a low yield strength, high work-hardening rate and high values of *n* and elongation.

The discovery of dual-phase steels is attributed to Rashid,[14] who found that a mixture of ferrite and martensite could be produced in a 0.15% CNbV steel by annealing in the intercritical, two-phase ferrite plus austenite region, between Ac$_1$ and Ac$_3$. By holding in this two-phase region, carbon can diffuse from the ferrite to the more accommodating austenite regions, raising the carbon content of the austenite to a level significantly higher than that of the nominal base composition. In turn, this increases the hardenability of the austenite, allowing martensite to form on cooling to ambient temperature.

It has been shown that the tensile strength of dual-phase steels is directly proportional to the martensite content and that the materials can be considered as mechanical mixtures of soft ferrite and hard martensite. Martensite contents of up to 15% are typical but dual-phase steels can develop tensile strengths in excess of 800 N/mm^2. Whereas they exhibit high *n* values, the deep-drawing characteristics are relatively poor with $\bar{r}$ values of around 1.0.

These steels are normally produced in both hot-rolled and cold-reduced gauges via continuous-annealing furnaces where the rapid cooling rates play an important part in promoting the formation of martensite on cooling from the intercritical annealing temperature. Even so, additions of silicon, manganese and chromium are sometimes incorporated in dual-phase steels in order to provide sufficient hardenability to ensure the formation of martensite. Dual-phase steels therefore tend to be expensive and this has undoubtedly inhibited their large-scale exploitation.

Bake-hardened steels

As discussed later, major problems were encountered with the introduction of higher strength steels to the automotive industry but Japanese steelmakers addressed these problems with a very novel metallurgical approach. The term *bake hardening* is derived from the paint baking or curing process that is applied to automobiles after the cold-forming and painting operations and, typically, it involves a treatment at 170°C for 20 minutes. During the cold forming of an automotive outer body panel, the strip undergoes a strain of the order of 2% and this results in an increase in yield strength of about 40 N/mm². Given a suitable composition and prior treatment, free interstitial carbon will precipitate on the dislocations generated in the cold-forming process during the bake-hardening treatment. In effect, bake hardening is a type of strain aging process which ordinarily is detrimental due to the accompanying loss in ductility but which can be used to considerable advantage after forming operations to provide a strengthening increment of about 50 N/mm².

Bake-hardening (BH) steels are supplied in the cold-reduced condition with maximum yield strength values of 250 N/mm² in order to overcome the problems associated with conventional high-strength steels. Whereas rimming steels, containing significant levels of free interstitial nitrogen, were capable of showing a substantial BH response, the change to aluminium-killed, continuously cast steels has necessitated the change to free carbon as the hardening agent. The BH response increases with increasing solute carbon, and provided the carbon content of the base steel is reduced to below 0.02%, sufficient carbon can be maintained in solid solution to provide a good response, even in batch-annealed material. In steels containing higher carbon contents, fine cementite particles are formed at the annealing temperature and act as nuclei for carbide precipitation on cooling to ambient temperature. In lower carbon steels, the elimination of these particles makes carbide precipitation very difficult and therefore significant amounts of carbon are retained in solid solution. The introduction of continuous annealing for strip offers greater scope for the production of BH steels, given the facility for varying the overaging time and temperature.

Small amounts of phosphorus and silicon are sometimes incorporated in these steels such that solid solution strengthening augments the increase in strength due to bake hardening.

Work-hardened steels

Although cold working produces a substantial increase in strength, this is accompanied by a major loss in ductility and therefore *dislocation strengthening* offers only limited potential in the area of high-strength strip steels. However, cold-worked materials are used in applications where the forming requirements are moderate. The ductility of cold-worked steels can be improved by means of heat treatments that produce recovery (*recovery annealed*) or partial recrystallization.

Transformation-strengthened steels

Depending upon the composition of the strip and the cooling rate from the austenitic region, a series of lower temperature transformation products can be produced, i.e. acicular ferrite, bainite or martensite. Yield strength values up to 1400 N/mm² can be produced by this route but again the materials have limited cold formability and softening can occur in the heat-affected zone after welding. Both work-hardened and transformation-strengthened steels are currently produced in very limited amounts.

Formability of high-strength strip steels

Each of the strengthening mechanisms outlined above reduces the ductility and cold-forming properties of the material but to a varying extent. The relationship between tensile strength and *n* value (stretch formability) for various types of high-strength steels is shown in Figure 2.14.[13] As indicated earlier, the dual-phase steels are characterized by a low yield strength and high rate of work hardening and these features explain the generation of high *n* values and good stretch-forming characteristics. This figure also underlines the limited forming characteristics of the transformation-strengthened and work-hardened steels.

In a similar vein, Figure 2.15[15] shows the relationship between tensile strength and *r̄* value (deep drawability) for various types of steels. This highlights the

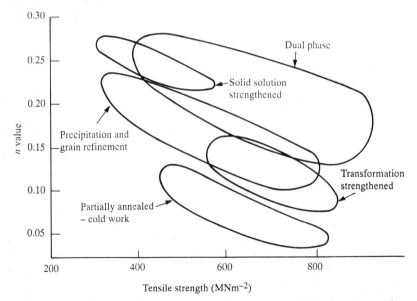

Figure 2.14   *Relationship between* n *value and tensile strength (After Dasarathy and Goodwin*[13]*)*

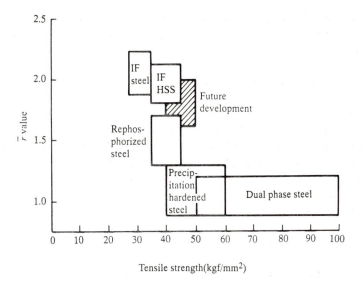

Figure 2.15 *Relationship between r value and tensile strength (After Takechi[15])*

excellent performance of the interstitial-free (IF and IF-HSS) steels and the relatively poor performance of the dual-phase steels in this particular straining mode. Although not featured in Figures 2.14 and 2.15, bake-hardening steels exhibit good stretching and drawing properties but, like the solid solution strengthened grades, their strength level is limited.

In addition to the type of strengthening mechanism employed, the cold-forming behaviour of steel strip is also affected by the level and type of non-metallic inclusions in the material. This aspect is particularly important in bending and hole expansion operations and in addition to making the steels to a low sulphur content, it might also be necessary to globularize the residual sulphides through the addition of calcium, titanium or the rare earth metals. Such treatments are dealt with in more detail in the chapter concerned with Low Carbon Structural Steels where inclusion-modifying agents of this type are added to improve the toughness of high-strength steels.

A particular problem that is encountered in the forming of high-strength strip is *springback* which is manifested by a change in shape when a component is taken out of the die. The departure from the required shape is due to the recovery of elastic strains in the component and the problem increases with increasing yield strength and decreasing thickness. The more generous radii that are frequently adopted in higher strength grades also intensify the degree of springback due to the reduction in plastic strain in the bend region. The general problem can sometimes be accommodated by over-bending such that the component recovers to the required shape after springback.

Standard specifications

BS 1449 has been in existence for many years, dealing with carbon and
carbon-manganese steels in the form of plate, sheet and strip. The last version
of this specification was issued in 1991 and Section 1.1 (General specification)
provides brief details of both hot rolled (HR) and cold rolled (CR) grades. A
summary of this information is given in Table 2.1 which indicates very broad
requirements for chemical composition in terms of maximum permitted levels
for carbon and manganese. There are no quantitative requirements for cold
formability in this standard, but under the heading *Quality*, an indication is
given as to the suitability of individual grades for particular forming
operations. Although this specification lacked quantitative details, it was the
tradition in the strip steel sector to supply on the basis of *fitness for purpose*
rather than rely solely on standard specifications. Grades CR1 to CR4 are all
made to carbon contents of about 0.04% with manganese contents of
0.20–0.25%. As such, the differences in formability are achieved by
variations in the processing conditions, the more formable grades being given
longer annealing times.

As indicated in the *Preface*, major changes have taken place in recent years
with the replacement of British standards by unified European standards. The
latter supersede British standards and BS EN 10130: 1991 (*Cold rolled carbon
steel flat products for cold forming*) is now the authoritative specification for
steel strip with specific requirements for cold formability. An extract from
this European standard is shown in Table 2.2. In addition to specifying tensile
properties, it should be noted that this new European standard also includes
requirements for minimum values of *r* (plastic strain ratio) and *n* (strain
hardening exponent).

Table 2.1 *BS 1449: 1991 Section 1.1 (General specification)*

Grade	Rolled condition	Quality	C% max.	Mn% max.
1	HR CR	EDDAK	0.08	0.45
2	HR CR	EDD	0.08	0.45
3	HR CR	DD	0.1	0.5
4	HR CR	D or F	0.12	0.6
14	HR	FL	0.15	0.6
15	HR	C	0.2	0.9

Key: EDDAK, extra deep drawing, Al-killed; EDD, extra deep drawing; DD, deep
   drawing; D or F, drawing or forming; FL, flanging; C, commercial.

Table 2.2 *BS EN 10130: 1991 Cold rolled carbon steel flat products for cold forming*

Grade	Definition and classification according to EN 10020	Deoxidation	Validity of mechanical properties [1]	Surface appearance	Absence of stretcher strain marks	R_e N/mm²	R_m N/mm² [2]	A_{80} % min. [3]	r_{90} min. [4,5]	n_{90} min. [4]	C	P	S	Mn	Ti
											\multicolumn{5}{Chemical composition (ladle analysis % max.)}				
Fe P01[6]	Non alloy quality steel[7]	Manufacturer's discretion	– / –	A / B	– / 3 months	–/280[8,10]	270/410	28			0,12	0,045	0,045	0,60	
Fe P03	Non alloy quality steel[7]	Fully killed	6 months / 6 months	A / B	6 months / 6 months	–/240[8]	270/370	34	1,3		0,10	0,035	0,035	0,45	
Fe P04	Non alloy quality steel[7]	Fully killed	6 months / 6 months	A / B	6 months / 6 months	–/210[8]	270/350	38	1,6	0,180	0,08	0,030	0,030	0,40	
Fe P05	Non alloy quality steel[7]	Fully killed	6 months / 6 months	A / B	6 months / 6 months	–/180[8]	270/330	40	1,9	0,200	0,06	0,025	0,025	0,35	
Fe P06	Alloy quality steel	Fully killed	6 months / 6 months	A / B	no limit / no limit	–/180[9]	270/350	38	$\bar{r}$ min.[4,5] 1,8	$\bar{n}$ min.[4] 0,220	0,02	0,020	0,020	0,25[11]	0,3

Notes:

1 The mechanical properties apply only to skin-passed products.

2 The values of yield stress are the 0,2% proof stress for products which do not present a definite yield point and the lower yield stress R_{eL} for the others. When the thickness is less than or equal to 0,7 mm and greater than 0,5 mm the value for yield stress is increased by 20 N/mm². For thicknesses less than or equal to 0,5 mm the value is increased by 40 N/mm².

3 When the thickness is less than or equal to 0,7 mm and greater than 0,5 mm the minimum value for elongation is reduced by 2 units. For thickness less than or equal to 0,5 mm the minimum value is reduced by 4 units.

4 The values of r_{90} and n_{90} or $\bar{r}$ and $\bar{n}$ (see annexes A and B) only apply to products of thickness equal to or greater than 0,5 mm.

5 When the thickness is over 2 mm the value for r_{90} or $\bar{r}$ is reduced by 0,2.

6 It is recommended that products in grade Fe P01 should be formed within 6 weeks from the time of their availability.

7 Unless otherwise agreed at the time of the enquiry and order Fe P01, Fe P03, Fe P04 and Fe P05 may be supplied as alloy steels (for example with boron or titanium).

8 For design purposes the lower limit of R_e for grade Fe P01, Fe P03, Fe P04 and Fe P05 may be assumed to be 140 N/mm².

9 For design purposes the lower limit of R_e for grade Fe P06 may be assumed to be 120 N/mm².

10 The upper limit of R_e of 280 N/mm² for grade Fe P01 is valid only for 8 days from the time of the availability of the product.

11 Titanium may be replaced by niobium. Carbon and nitrogen shall be completely bound.

After BS EN 10130: 1991.

Zinc-coated steels

Coating with zinc (*galvanizing*) is one of the most widely used and cost-effective means of protecting mild steel against atmospheric corrosion. Zinc itself has good resistance to corrosion through the formation of protective surface films of oxides and carbonates but zinc coatings protect steel in two ways:

1. By forming a physical barrier between the steel substrate and the environment.
2. By providing *galvanic* or *sacrificial* protection by virtue of the fact that zinc is more electronegative than iron in the electrochemical series. This effect is illustrated schematically in Figure 2.16.

Production methods

Zinc coatings are applied continuously by the *hot dip galvanizing* and *electrogalvanizing* processes, and in order to appreciate the difference in properties between the products, the two processes will be briefly described.

Hot dip galvanizing (HDG)

The various stages of a modern HDG line are shown schematically in Figure 2.17. Cold-rolled strip is first welded to the trailing edge of the previous coil and enters an accumulator or storage station. This enables the process section of the line to function whilst the cutting and joining operations are being carried out.

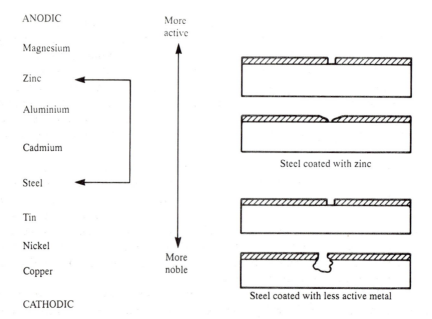

Figure 2.16   *Galvanic protection of steel by zinc*

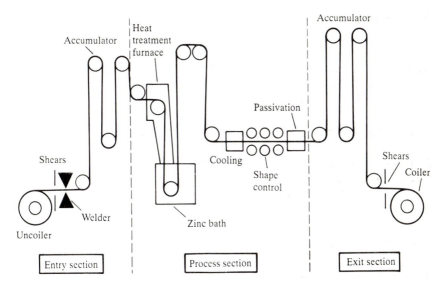

Figure 2.17 *Hot dip galvanizing line*

The strip then passes into a heat treatment furnace with an inert atmosphere. Initially, the strip is heated to a temperature of about 680°C to burn off rolling oil and the material then enters the annealing section of the furnace. In this section, the strip attains a temperature of about 730°C and the cold-rolled strip recrystallizes. Still under a protective atmosphere, the strip is cooled rapidly by means of a gas stream, before entering the zinc bath which is maintained at a temperature of 465°C.

Small amounts of aluminium, typically 0.15%, are added to the bath to restrict the thickness of the Zn–Fe layer and thus provide an adherent coating that can accommodate major strains in subsequent forming operations. The aluminium reacts preferentially with the steel as it enters the galvanizing bath, forming a thin layer of Fe–Al compounds ($FeAl_3$, Fe_2Al_5), and thereby retards the reaction between iron and zinc. The lead content of the bath affects the zinc crystal size or *spangle* on the surface of the strip, and for applications requiring a minimum or spangle-free appearance, the lead level of the bath is held below 0.15%.

After the coating operation, the strip passes through a series of *gas-knives* which use air or steam to control the amount of zinc on the surface of the steel. Depending on the end application, the strip may then be temper rolled or stretcher levelled to provide shape control or to eliminate the formation of strain markings (*Lüders lines*) during cold forming. The final stage in the process is generally a passivation treatment in which the strip is sprayed with a solution of chromic acid. This inhibits the formation of a corrosion product known as *white rust* which will be discussed later.

The HDG process is also used to produce *galvanneal* strip in which the plain zinc coating of the traditional galvanized product is converted to a Zn–Fe alloy with an approximate composition of 90% Zn, 10% Fe. Galvanneal strip is produced by passing the HDG strip into a heated chamber directly after leaving the coating bath. This accelerates the diffusion process and the zinc coating is transformed into a layer of Zn–Fe intermetallic compounds. The formation of this layer is critically dependent on holding time in the diffusion chamber and the coating weights produced on galvanneal steel are at the bottom of the range of those produced on plain zinc coatings.

Galvanneal coatings have a matt grey appearance of low reflectivity and have the following characteristics:

1. Similar corrosion resistance to plain zinc coatings.
2. Good paint adherence.
3. Better resistance welding performance than plain zinc coatings.

Galvanneal strip is also known by the letters IZ, which refer to the iron–zinc coating.

Electro-galvanizing

In contrast to the HDG process, electro-galvanizing is carried out at or near ambient temperature, the feedstock having been annealed in the conventional manner. Prior to the coating operation, the strip is first passed through a series of chemical or electrolytic cleaning baths to remove dirt or oil. The plating solution is made up primarily of zinc sulphate and zinc is supplied to the electrolyte by means of zinc anodes. After plating, the strip is passed through chromate baths for passivation.

Corrosion resistance

The life of a zinc coating on steel is roughly proportional to its thickness and therefore galvanized strip is produced in a range of coating thicknesses to satisfy different conditions and end uses. In the hot dip product, the amount of zinc on the steel is expressed in g/m^2, the figures after the letter G (plain zinc) and IZ (iron–zinc alloy) in Table 2.3 indicating the minimum mass per unit area, including the coatings on both sides of the strip.

Table 2.3

Light coating	G100	Where corrosion conditions are mild or where
	G200	forming requirements preclude heavier coatings
Medium coating	G275	Standard coating
Heavy-duty coating	G350	For longer life requirements
	G450	
	G600	
Iron–zinc coating	IZ100	Fe–Zn coatings for good painting and welding
	IZ180	characteristics

In electro-galvanized strip, the coatings are significantly lighter, as shown in Table 2.4.

Table 2.4

Grade	Nominal coating thickness per surface (μm)	Nominal coating mass per surface (g/m^2)	Minimum coating mass per surface (g/m^2)
EZ 10/10	1.0	7	4
EZ 25/25	2.5	18	12
EZ 50/50	5.0	36	28
EZ 75/75	7.5	54	47

In the above table, it should be noted that the coating mass relates to a single surface rather than the combined mass for the two surfaces, which is the convention used for the hot dip product. Electro-galvanized material can also be supplied with single-sided (e.g. EZ 25/00) or differential (e.g. EZ 75/25) coatings.

As with other steel products, the atmospheric corrosion resistance of zinc-coated steel is affected greatly by the nature of the environment and it is common to differentiate between the performance in rural, industrial and marine locations. The performance of hot dip galvanized steel, in terms of life to first maintenance, is illustrated in Table 2.5. This indicates the following features:

Table 2.5   *Atmospheric corrosion performance of hot dip galvanized steel*

Hot dip zinc coating to BS 2989		Life to first maintenance in the United Kingdom type of environment				
		External			Internal	
Description	Mass (g/m^2) including both sides[a]	Coastal	Industrial and urban	Suburban and rural	Wet and possibly polluted	Dry and unpolluted
Type G275	275	S	S	M	XX	VL
Type G350	350	S	S	M	XX	VL
Type G450	450	M	S	L	XX	VL
Type G600	600	L	M	VL	XX	VL

The recommendations for life to first maintanance are based on the premise that the strength and integrity of the sheet would diminish with time after the period specified unless maintenance was carried out.

S, short 2–5 years; M, medium 5–10 years; L, long 10–20 years; VL, very long 20–50 years; XX, further protection with paint is recommended for this type of application.

[a] Minimum average triple spot.

After Galvatite Technical Manual
(British Steel Strip Products).

1. The detrimental effect of sulphur dioxide (industrial) and chloride ions (marine) in the atmosphere.
2. The beneficial effect of increasing the thickness of the zinc coatings.
3. The need for paint protection in internal applications if the conditions are wet and polluted.

When zinc-coated products are stored under wet conditions or where condensation can occur, *white rust* can form on the surface. This is due to the formation of zinc carbonate and detracts from the appearance of smooth zinc coatings. However, although it may form in large amounts, the appearance of white rust does not necessarily indicate severe degradation of the zinc coating and it will generally convert to a protective layer. As indicated earlier, the problem can be avoided by immersing galvanized steel in chromic acid solutions to passivate the surface.

Care must be taken when fixing zinc-coated steels to other metals in order to avoid galvanic corrosion. In particular, copper or brass should not be coupled directly to galvanized steel since the coating can fail rapidly in wet and polluted atmospheres.

In aqueous media, the performance of zinc-coated steels is affected by a number of factors:

1. *pH of solution* – the corrosion rate is generally low in the pH range 6–12 but can be rapid outside this range.
2. *Hardness of water* – hard water precipitates carbonates on zinc surfaces which reduce the rate of corrosion.
3. *Water temperature* – whereas zinc provides sacrificial protection to the steel substrate at ambient temperature, the reverse situation occurs at temperatures above 60–70°C. Therefore, zinc-coated steels should not be used in hot-water systems.
4. *Chlorides* – soluble chlorides can produce rapid attack, even within the pH range 6–12.

Cold-forming behaviour

The cold-forming behaviour of zinc-coated strip is governed primarily by the forming characteristics of the substrate. As indicated earlier, in hot dip galvanizing, cold-reduced strip is subjected to rapid heating and cooling cycles, similar to those employed in continuous annealing but generally without an overaging stage. Therefore, the formability of hot dip galvanized steel is poorer than that of uncoated, batch-annealed material. However, the formability of the hot dip product can be improved through the use of modified annealing cycles but the special IF (interstitial-free) steels can also be employed as a substrate for highly formable, hot dip galvanized strip. The IF steels were referred to earlier on p. 25 and owe their good formability to the elimination of carbon and nitrogen from solid solution.

As described earlier, electro-galvanizing takes place at near-ambient temperature and therefore the coating operation has no effect on the forming behaviour or mechanical properties of the substrate. Electro-galvanized strip can therefore be supplied to the CR1, CR2, CR3 and CR4 grades specified in BS 1449: Part 1 and electro-galvanizing can also be applied without detriment to rephosphorized and other grades of high-strength steels.

Standard specifications

Hot dip zinc strip steels were covered previously in BS 2989: 1982 but this specification has now been superseded by European specification BS EN 10142: 1991 – *Continuously hot dip zinc coated low carbon steel sheet and strip for cold forming*. This specification involves a complex but logical system of designation which is built up in the following manner:

(a)   The number of the standard – BS EN 10142.
(b)   Full designation of the steel grades (e.g. Fe PO3 G – see Table 2.6).
(c)   A letter indicating type of coating:
      Z   Zinc coating
      ZF   Zinc-iron alloy coating
(d)   A number denoting mass of coating (e.g. 275 = 275 g/m^2, including both sides).
(e)   A letter denoting coating finish:
      N   Normal spangle
      M   Minimized spangle
      R   Regular zinc-iron alloy coating
(f)   A letter denoting surface quality:
      A   As coated surface
      B   Improved surface
      C   Best quality surface
(g)   A letter denoting surface treatment:
      C   Chemical passivation
      O   Oiling
      CO Chemical passivation and oiling
      U   Untreated

An example of a designation built up in this manner would be:

BS EN 10142 – Fe PO3 G Z 275 NA – C

As illustrated in Table 2.6, this European specification contains four basic steel grades listed in order of increasing suitability for cold forming:

Fe PO2 G   bending and profiling quality
Fe PO3 G   drawing quality
Fe PO5 G   deep drawing quality
Fe PO6 G   special deep drawing quality

These basic grades can be specified with a variety of specific minimum coating masses (both zinc and zinc-iron coatings), as illustrated in Table 2.7.

Table 2.6   *BS EN 10142: 1990 Continuously hot dip zinc coated low carbon steel sheet and strip for cold forming. Steel grades and mechanical properties*

Steel grade	Yield strength R_e, max.[1,2] N/mm^2	Tensile strength R_m, max.[2] N/mm^2	Elongation A_{80}, min.[3] %
Fe PO2 G	–	500	22
Fe PO3 G	300[4]	420	26
Fe PO5 G	260	380	30
Fe PO6 G	220	350	36

Notes:
1 The yield strength values apply to the 0,2% proof stress if the yield point is not pronounced, otherwise to the lower yield point (R_{eL}).
2 For all steel grades a minimum value of 140 N/mm^2 for the yield strength (R_e) and of 270 N/mm^2 for the tensile strength (R_m) may be expected.
3 For product thicknesses $\leq$ 0,7 mm (including zinc coating) the minimum elongation values (A_{80}) shall be reduced by 2 units.
4 This value applies to skin passed products only (surface qualities B and C).

After BS EN 10142: 1990.

Table 2.7   *BS EN 10142: 1990 Continuously hot dip zinc coated low carbon steel sheet and strip for cold forming. Coating masses*

Coating designation	Minimum coating mass g/m^2, including both surfaces[1]	
	Triple spot test	Single spot test
Z100, ZF100	100	85
Z140, ZF140	140	120
Z200	200	170
Z225	225	195
Z275	275	235
Z350	350	300
Z450	450	385
Z600	600	510

Note:
1 The coating mass of 100 g/m^2 (including both surfaces) corresponds to a coating thickness of approximately 7,1 µm per surface.

After BS EN 10142: 1990.

Zinc alloy coatings

Whereas IZ strip has a Zn–Fe coating which is produced by diffusion from a plain zinc layer, zinc alloy coatings are also deposited directly onto the steel.

55% Al–Zn

In 1972, the Bethlehem Steel Corporation introduced steel strip with a coating of 55% Al 43.5% Zn 1.5% Si under the name *Galvalume*. Since that time, the manufacture of the product has been licensed throughout the world and it is marketed in the UK under the name *Zalutite*. The 55% Al–Zn coating is applied by the hot dip process route in a similar manner to that used for the conventional, plain zinc coatings. On cooling from the coating bath, an aluminium-rich phase is the first to solidify and makes up about 80% of the volume of the coating. The remainder is made up of an interdendritic, zinc-rich phase and an Al/Fe/Zn/Si intermetallic compound bonds the coating to the steel substrate, providing further resistance to corrosion. Silicon is added to the Al–Zn alloy in order to restrict the growth of the brittle intermetallic layer.

The coating was developed specifically to provide an improved corrosion performance compared with plain zinc coatings and benefit is derived from the separate effects of zinc and aluminium. In the initial stages of corrosion, attack takes place preferentially on the zinc-rich phase until its corrosion products stifle further activity in these areas. However, as well as acting as a barrier to the transport of corrodents, the zinc also provides sacrificial protection at cut edges and areas of damage. As the zinc-rich phase is leached away, corrosion protection is provided by the aluminium-rich phase which forms protective films of oxides and hydroxides on the surface of the material. As indicated in Table 2.8, the 55% Al–Zn layer provides between two and four times the life of

Table 2.8 *Comparative corrosion losses as a decrease in thickness (micrometres) for 55% Al–Zn alloy coated and hot dip zinc-coated steel strip at Australian test sites*

Site	Years exposed	(A) Hot dip zinc-coated (μm)	(B) 55% Al–Zn alloy (μm)	Ratio A/B
Severe marine	2.5[a]	16.8	5.2	3.2
Industrial marine	7	10.5	4.7	2.2
Industrial	7	9.8	3.4	2.9
Marine	4	5.9	1.4	4.2
Rural	4	1.4	0.8	1.8

[a] Exposure was discontinued after $2\frac{1}{2}$ years because all the coating on the groundward surface of the hot dip zinc-coated sample had been lost by that time. No rust on remaining 55% Al–Zn alloy-coated samples still on exposure after seven years.
After *Zalutite Technical Manual*
(British Steel Products).

conventional zinc coatings, depending upon the nature of the environment.

The material is produced with standard coating masses of 150 (GA150) and 185 (GA185) g/m^2, including both surfaces. These values equate to coating thicknesses of 20 and 25 µm respectively, on each surface. The forming properties of the material are generally similar to those of continuously annealed, hot dip zinc-coated steel. However, it has an increased tendency to spring back and lacks the self-lubricating properties of hot dip zinc coatings. The application of an effective lubricant is therefore essential.

The 55Al–Zn coating has a smooth, silvery appearance with a very fine spangle and is said to be attractive in the unpainted condition. However, the material is also supplied in the factory-painted condition and painting is recommended for use in severe marine and very corrosive environments. The coating exhibits good resistance to heat and can withstand discoloration in air at temperatures up to 310°C, whereas a limit of 230°C is prescribed for hot dip zinc coatings.

95% Zn 5% Al

A coating of this composition was first introduced in 1982 and is marketed world-wide under the name *Galfan*. The coating is again deposited by the hot dip process and, in addition to good corrosion properties, it is claimed to have particularly good forming characteristics. The structure of the coating is dependent on the rate of cooling from the bath and cooling rates greater than 20°C/s result in the formation of a fine eutectic of zinc-rich and aluminium-rich phases. Slower cooling rates result in the separation of a primary zinc-rich phase.

87% Zn 13% Ni

Both forming and welding problems are introduced with the application of heavy zinc coatings and therefore there is a demand for coatings that will provide good corrosion resistance, even when applied in thin layers. One approach to this problem is the incorporation of nickel into zinc-based coatings, and a coating containing 87% Zn 13% Ni is now being used for automotive applications. The coating is applied by the electro-plating route and the material is marketed in the UK under the name *Nizec*.

Organic-coated steels

During the 1980s there was a dramatic growth in the production of pre-finished strip, coated with various types of paint or plastic. These coatings are available in a wide range of colours and textures and are used to advantage where corrosion resistance and a decorative appearance are of major concern. The coatings are applied to a range of steel substrates, but generally to zinc-coated strip, and are formulated specifically for various manufacturing requirements or end uses. Coatings are available that provide long life in external applications, good deep-drawing characteristics and resistance to heat or chemical attack.

The organic coatings are applied continuously as a liquid film or as a laminate which is bonded to the substrate with an adhesive. In either case, the steel substrate is thoroughly cleaned in a multi-stage process to ensure uniform and optimum adherence of the coating. In the liquid film route, the substrate is first coated with a primer, which is cured, and the top coating is then applied by the reverse-roll method. Finally, the topcoat is cured in a finishing oven. Where embossing is required, typically on PVC coatings of 200 μm, a patterned steel roll is applied to the hot PVC as it emerges from the finishing oven. The material is then immediately quenched in water in order to 'freeze' in the embossed texture. Coil-coated strip is marketed in the UK under the name *Colorcoat*.

In the laminate coating, an adhesive and backing coat are applied by roller-coating and the strip is passed through an oven to activate the adhesive and cure the backing coat. The coating film, generally PVC, is bonded to the steel and cooled immediately by water quenching. The material is marketed in the UK under the name *Stelvetite* and is intended for internal applications.

Organic-coated strip is specified in Euronorm 169-85. The following information on the characteristics of the more important coatings has been derived from Annex A of Euronorm 169-85 and a British Steel publication:[16]

- *PVC Plastisol* (200 μm) – a plasticizer-bearing coating with very good flexibility. Can be drawn and formed easily. Suitable for embossing for decorative purposes and can be used in internal and external applications. *Typical applications* – roofing and cladding on buildings, curtain walling, furniture, vehicle fascia panels, garage doors.
- *PVC Organosol* (50 μm) – a coating with good flexibility and specially recommended for deep-drawn parts. Not recommended for exterior use. *Typical applications* – electric light fittings, cable trunking, deep-drawn parts.
- *Acrylic* (25 μm) – unplasticized coating with good flexibility. Suitable for continuous operation at temperatures up to 120°C. Good resistance to chemical attack.
- *Epoxy* (5–15 μm) – hard, chemical-resistant coating with good flexibility. *Typical applications* – used extensively as a primer for two-coat systems. Good adhesion to polyurethane foam. Not recommended for external use.
- *Polyesters* (25 μm) – widely applied coatings with good flexibility and suitable for continuous exposure at temperatures up to 120°C. Various formulations are available that offer good deep-drawing properties, resistance to chemical attack and that are suitable for exterior use. *Typical applications* – consumer durables, deep-drawn components, building components.
- *PVF2* (27 μm) – a coating with good flexibility and highly resistant to chemicals and solvent attack. Highly resistant to weather and particularly suitable for exterior use. *Typical applications* – building components.

As indicated above, many of these coatings exhibit good formability and are amenable to forming operations such as press braking and folding, roll forming and deep drawing. However, the flexibility of the coatings varies with ambient temperature and, to minimize the risk of cracking during forming, it is recommended that the liquid film and laminate-coated materials should be allowed to attain minimum temperatures of 16 and 20°C respectively.[17]

   The resistance welding of organic-coated steel is not possible by conventional methods because the coating prevents the flow of current across the electrodes. On the other hand, resistance welding is possible if the coating is removed locally, or if special capping pieces are inserted to melt the coating.[17] However, mechanical joining techniques have been developed specifically for coated steels and adhesives are also available that are suitable for particular types of coating.

Steel prices

The basis prices of standard hot-rolled and cold-reduced strip grades are shown in Figure 2.18. These prices were effective on 1 March 1994 and it must be

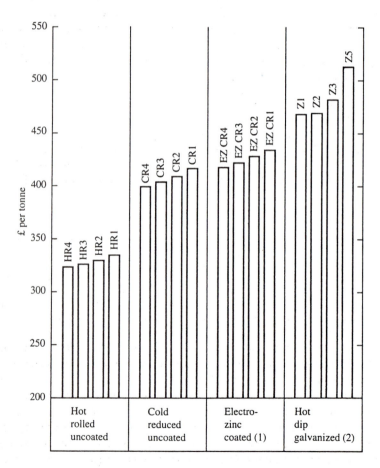

Figure 2.18   *Basis prices of strip mill products: (1) zinc coating 2.5 μm both sides; (2) zinc coating 275 g/m², including both sides; as of March 1994*

borne in mind that steel prices are adjusted periodically. This figure shows that a substantial differentiation is made between the hot-rolled and cold-reduced conditions, reflecting the additional processing costs associated with the latter. However, the cost differential within these two conditions with respect to the various formability criteria is relatively small.

Figure 2.18 also includes the basis prices for zinc-coated steels. This indicates that electro-galvanized strip costs about £15/tonne more than uncoated, cold-reduced steel. However, the cost differential is substantially more in the case of hot dip galvanized strip, which has a heavier zinc coating than the electro-galvanized product.

Tinplate

Tinplate enjoys a pre-eminent position in the packaging industry, particularly in relation to cans for food and beverages. Substantial amounts of tinplate are also used in aerosol containers and for the packaging of paints and oils. This section reviews briefly the manufacturing process for tinplate and the basic procedures employed in canmaking. These specific aspects, together with other important technical details relating to tinplate, are covered in a very definitive and comprehensive publication by Morgan.[18]

Method of manufacture

Virtually all tin mill products in the UK are produced from continuously cast steel to the chemical composition shown in Table 2.9. Where special corrosion resistance is required, a steel with 0.015% P max. and 0.06% Cu max. (Type L) is supplied.

Table 2.9

				Weight % max. (Type MR)					
C	*Mn*	*P*	*S*	*Si*	*Cu*	*Ni*	*Cr*	*Mo*	*Al*
0.13	0.6	0.02	0.05	0.03	0.2	0.15	0.1	0.05	0.1

The starting point in the manufacture of tinplate is hot-rolled strip with a typical thickness of about 2 mm. This is pickled to remove the scale formed during hot rolling and the material is then cold rolled to the required thickness in either a single-stage or a two-stage operation. For *single-reduced* tinplate, the cold-rolling reduction is of the order of 90% and the material is available in the thickness range 0.16–0.6 mm. After cold rolling, the strip is softened by either batch or continuous annealing in order to restore ductility but the annealed material is then given a light cold reduction, termed *temper rolling*, before the tinning operation. Temper rolling improves the surface finish and flatness of the strip and also provides the required mechanical properties for particular applications. Thus single-reduced tinplate is available in a range of tempers with *different strengths* and this is illustrated in Table 2.10.

In *double-reduced* tinplate, cold-rolled and annealed strip is subjected to a second cold reduction of 30–40%. No further annealing is undertaken and the material is substantially work hardened, exhibiting a marked directionality in properties. The bulk of double-reduced tinplate is produced to the thickness range 0.16–0.18 mm but is available in the range 0.13–0.27 mm. Typical mechanical properties are shown in Table 2.11. By virtue of its higher strength, double-reduced tinplate provides the facility to reduce the cost of a can or other components by decreasing the thickness of the material without loss of rigidity.

Table 2.10   *Tinplate – single-reduced tempers*

Temper number	Euronorm/ISO temper designation	Target hardness HR 30T	Typical application
T1	T50	52 max.	Mainly used for components which make the maximum demand on the formability of the steel base, e.g. deep-drawn containers, bakeware, puddings basins, oil filter bodies
T2	T52	52	Typically used for forming operations which are less severe than above, e.g. shallow-drawn bakeware, rectangular caps, cushion rings for paint cans
T3	T57	57	Used for a wide range of applications where moderate formability is required, e.g. shallow-drawn components, can bodies
T4	T61	61	Typically used for ends, bodies, stampings where a stiffer and stronger product is required
T5	T65	65	This is the strongest product available in the conventional temper range and is typically utilized for stiff ends and bodies

After British Steel Tinplate – Product Range

Table 2.11   *Tinplate – double-reduced tempers*

Designation		Target hardness HR 30T	Typical applications
New	Previous		
DR550	DR8	73	Round can bodies and can ends
DR620	DR9	76	Round can bodies and can ends
DR660	DR9M	77	Beer and carbonated beverage can ends

Following the cold-rolling sequences described above, the strip is coated with tin. A small amount of material is still coated by hot dip tinning, similar to that described earlier for galvanizing, but the *Ferrostan* process involving the electro-deposition of tin now accounts for the bulk of tinplate production. Prior to plating, the strip is thoroughly cleaned in electrolytic pickling and degreasing units, followed by washing. The Ferrostan process uses a bath of acid stannous sulphate and tin with a purity of not less than 99.85%. After plating, the coating is flow-melted by resistance heating to a temperature above the melting point of tin (232°C), e.g. 260–270°C, followed by water quenching. This treatment produces a reflective surface and also results in the formation of an iron–tin compound ($FeSn_2$) which plays an important role in the corrosion resistance and soldering characteristics of the material. The product is then passivated by immersion in a dichromate solution which deposits a very thin film of chromium on the surface. After passivation, a thin film of oil is applied in order to preserve the surface from attack and also to enhance the lubrication properties in subsequent handling operations. The oil applied, dioctyl sebacate, is a synthetic organic oil and is acceptable for use in food packaging.

Electrolytic tinplate can be produced with equal or differential coatings on each surface, the former carrying the prefix *E* in the designation and the latter having the prefix *D*. The mass on each surface is expressed in g/m^2 and thus E 2.8/2.8 has 2.8 grams of tin per square metre on each surface, giving a total of 5.6 g/m^2. A tin coating of 5.6/5.6 g/m^2 is equivalent to a coating thickness of 0.75 μm per surface. The normal range of equally coated products is from E 1.4/1.4 to E 11.2/11.2. However, low-tin coatings have been developed down to 1.0/1.0 g/m^2. In differentially coated products, the normal range is from D 2.0/1.0 to D 11.2/5.6.

Tinplate manufacturers also produce other products such as uncoated and oiled sheet (*blackplate*) and material coated with metallic chromium. According to Morgan,[18] blackplate has not achieved significant usage in canmaking operations due to problems in providing adequate resistance to corrosion by lacquering techniques. However, electrolytic chromium/chromium oxide-type coatings have enjoyed greater success and were developed primarily because of the extremely variable price of tin. The coating is duplex and consists of about 80% metallic chromium adjacent to the steel substrate and 20% hydrated chromic oxide/hydroxide in a layer above. It is recommended that this type of coating is lacquered to provide added surface protection and to enhance fabrication.

Canmaking processes

Three basic procedures are employed:

1. Three-piece can manufacture.
2. Drawing and wall-ironing.
3. Draw and redraw.

Three-piece cans consist of a welded cylindrical body and two ends, one of which (the base) will be attached to the body of the can by the canmaker and the other will be applied after filling. The cylinder is rolled into shape from flat, pre-lacquered, rectangular blanks and the two edges are joined by electrical welding. A further coat of lacquer is then applied to the weld seam. The can ends are pressed from circular blanks in an operation that requires a high degree of precision. The ends are contoured with a series of expansion rings so that they can support the internal pressure through tensile rather than bending stresses. The rims of the ends are also carefully stamped and curled so as to accept a sealing compound which forms an airtight seal with the body of the can. Cans of this type are used for most human foods and also for paints, oils and chemicals.

The *drawing and wall-ironing* (DWI) process is more modern than three-piece canmaking and is illustrated schematically in Figure 2.19. Thus DWI eliminates the need for a welding operation and relies solely on presswork. Morgan cites the following advantages for DWI over three-piece cans:

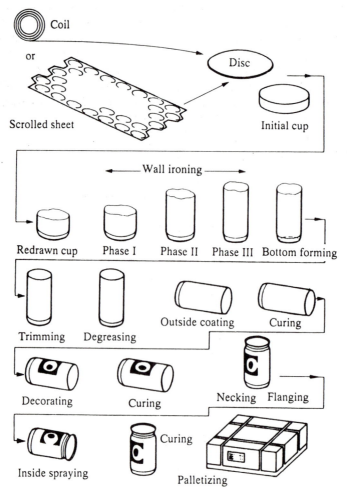

Figure 2.19   *Drawing and wall-ironing operation sequence (After Morgan[18])*

1. More effective double seaming of the top end to body, because of the absence of the sensitive side seam junction (particularly valuable for processed foods and carbonated beverages).
2. Significantly lower metal usage and cost.
3. More attractive appearance due to the absence of a side seam.

Following the initial drawing operation, the cup is redrawn to final can diameter. During *wall-ironing*, the can may pass through a series of dies which produce a substantial reduction in wall thickness and a complementary increase in body height. The principle of wall-ironing is shown in Figure 2.20. Following trimming, the can is degreased prior to applying decoration to the outside and lacquer to the inside. DWI cans are used for beer and soft drinks, for pet foods and some human foods. Aluminium is a major competitor to tinplate in DWI cans for these products, although tinplate enjoys a larger share of the UK market at this time.

In the *draw and redraw* (DRD) process, the initially drawn cup is redrawn to one of smaller diameter and greater height in one or two redrawing operations. Like the DWI process, it eliminates the side seam but the maximum height/diameter ratio that can be produced by multiple drawing is less than that achievable by DWI. However, DRD has a major advantage over DWI in that pre-lacquered tinplate can be used, thus eliminating the costly cleaning and spray-lacquering operations at the end of the process.[18] DRD cans are used for pet foods and some human foods, including baby foods.

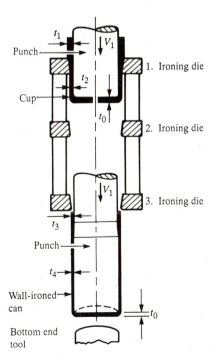

Figure 2.20 *Wall-ironing with three ironing dies (After Morgan[18])*

Canmaking via DWI and DRD takes place at high speed and involves severe plastic strain. The steel therefore needs to be of the highest quality and a very low level of non-metallic inclusions is essential to the efficient operation of these processes. Gauge control is also important. The DRD process requires higher $\bar{r}$ values so that extensive drawing can take place. However, care must be taken to avoid an excessively large ferrite grain size which can give rise to an orange peel effect and a poor surface for lacquering.

Resistance spot welding

Resistance welding is produced by generating heat from the resistance to the flow of electric current through work pieces that are held together by the force of clamping electrodes. There are various forms of resistance welding, including spot, seam and projection welding, all of which are used extensively in the welding of sheet materials. However, *spot welding* has been selected for description in view of its relative simplicity and its wide range of application in low-carbon strip steel.

Spot welding is generally used for joining steel strip of up to 3 mm thick, although thicknesses up to 10 mm can be welded by this process. The size and shape of the welds are controlled primarily by the size and shape of the water-cooled electrodes. In general, the weld nugget should be oval in cross-section but should not extend completely to the outer surfaces. The electrodes exert a significant clamping force on the strip materials and the contacting surfaces are generally heated by pulses of low voltage (5–20 V), high current (5000–20 000 A), 50 cycle AC electricity. When sufficient melting is achieved, the current is switched off but the clamping force of the electrodes is maintained until the weld pool has completely solidified. The cycle is completed in a fraction of a second. Spot welding is also carried out with DC machines.

The integrity of a spot weld is judged by means of a *peel test* or *chisel test*, which separate the sheet materials after welding. In some cases, separation occurs through the weld and along the original interface of the materials (*interface failure*) or else the material tears around the weld nugget (*plug failure*). A full pull-out or plug failure is generally regarded as indicative of a good sound weld whereas interface fractures are associated with brittle and undersized welds. Tensile tests, using tensile-shear and cross-tension configurations, are also employed for evaluating the static strength of spot welds.

Weldability lobes

The heat generated during resistance heating can be expressed by the following equation:

$$H = I^2 R T$$

where H = heat in joules

         I = current in amps

R = resistance in ohms
T = time in seconds

Therefore, for a fixed weld time T_1, the size of the weld nugget will increase with current, according to the relationship shown in Figure 2.21(a). At a particular level of current a_1, the weld diameter will reach what is regarded as the minimum acceptable size, namely a diameter of at least $4\sqrt{t}$, where t is the sheet thickness. As the current is increased beyond this critical level, the size of the nugget will increase until the stage is reached when the weld pool breaks the surface, giving a condition which is termed *expulsion* or *splash*. This is achieved at current a_2 and this condition is regarded as unacceptable, irrespective of weld diameter. Therefore, for a given welding time, there exists a range of current that will produce acceptable welds, i.e. from that which just meets the minimum acceptable size criterion to that which just avoids the splash condition. This type of exercise can be repeated for different weld times and an acceptable range of current can be defined for each particular time. These combinations of current and time that produce acceptable welds are then expressed in the form of weldability *lobes*, as illustrated in Figure 2.21(b). Thus currents or times below the lower bound of the lobe produce welds that are below the minimum acceptable size and which generally exhibit interface failure. Conversely, the combination of currents and times above the lobe will lead to the splash condition.

The size of the lobe can be taken as a measure of the weldability of a material, large lobes indicating a greater tolerance to changes in production conditions.

Rephosphorized steels

Because of the interest in rephosphorized steels for body panels in the automotive industry, a substantial amount of work has been carried out on these steels in order to achieve satisfactory welds using conventional resistance welding equipment. Jones and Williams have published extensive reviews[19, 20] on the spot welding characteristics of these grades and their main conclusions are summarized below:

1. The weldability lobes of rephosphorized steels are sufficiently wide to present few production problems, the lobe widths being only slightly less than those obtained in plain carbon steels of similar thickness.
2. The acceptable range of current (*available current*) can be improved by using larger electrode forces and the optimum results are obtained by increasing these forces by at least 50% compared with comparable thicknesses of plain carbon steel.
3. The available current can also be increased significantly by using larger electrode tip sizes, e.g. the width of the lobe can be increased by 700–2000 A by increasing the tip diameter from 4.8 to 6.4 mm.

Alloying elements increase the resistivity of low-carbon ferritic steels and consequently less current is required to produce a weld of a given size in a

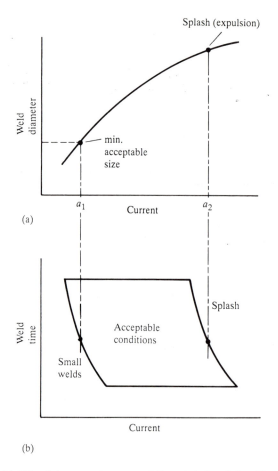

Figure 2.21   *Weldability lobes – resistance welding; (a) fixed time T_j; (b) for various times*

rephosphorized grade compared with its phosphorus-free counterpart. However, the available current range may be more important than the actual current level but, as indicated above, the adverse effect of phosphorus is minimal.

Both carbon and phosphorus can influence the fracture mode of spot welds and partial plug failures are the most common type of failure in peel tests on rephosphorized steels with a weld diameter of $5\sqrt{t}$. However, 100% plug failures can be obtained in rephosphorized steel of 0.67 mm thickness by increasing the weld size to greater than $7.5\sqrt{t}$.[19] In order to obtain an optimum balance between strength and weldability, the combined carbon plus phosphorus contents are generally limited to a maximum of 0.18%.

Precipitation-strengthened grades

Typical weldability lobes for niobium- and titanium-strengthened grades are shown in Figure 2.22.[19] In general, niobium-treated steels require a lower welding current to produce a given weld size but both niobium and titanium

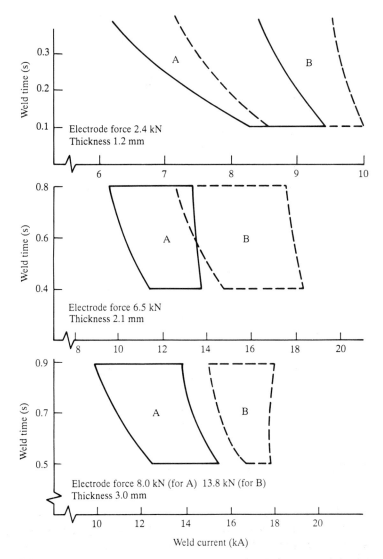

Figure 2.22 *Comparison of weldability lobes for niobium and titanium strengthened steel of thickness 1.2–3 mm: A = niobium; B = titanium (After Jones and Williams[19])*

lower the level of current required compared with unalloyed steels. As illustrated in Figure 2.22, the difference in behaviour between the niobium- and titanium-strengthened grades intensifies with increasing sheet thickness. In common with other higher strength steels, the precipitation-strengthened grades require higher electrode clamping forces than unalloyed, plain carbon steels.

Plug failures are generally obtained in these steels with sheet thicknesses up to 1.6 mm. However, to produce plug failure in sheet thicknesses greater than 1.6 mm, multiple impulse welding or a post-heat pulse must be included in the schedule.[19]

Zinc-coated steels

In relation to hot dip products, distinction must be made between plain zinc and IZ alloy coatings. Because of its lower contact resistance, higher currents are required for plain zinc-coated strip, such as G275, compared with equivalent thicknesses of uncoated mild steel. On average, the currents have to be increased by about 20%. On the other hand, IZ-coated strip has higher surface resistance than G275 and requires only a slight increase in current compared with uncoated steel. The weld times employed for zinc-coated steels are 50–100% longer than those used for uncoated steel and the electrode pressure must be in the region of 90 N/mm^2 and 150 N/mm^2 for IZ and G275 respectively, when using truncated cone-type electrodes. Electro-deposited zinc coatings require slightly higher welding currents for a given weld size than hot dip coatings, the absence of an alloy layer in the former producing a lower contact resistance. Higher welding currents are also required in electro-galvanized steel as the thickness of the coating is increased. On the other hand, the position and width of the weldability lobes for hot dip zinc-coated steels are insensitive to coating thickness.

The spot welding of zinc-coated steels is characterized by an electrode life which is shorter than that obtained with uncoated steels. This results from the fact that coated steels require higher welding currents and longer welding times, both of which increase the degree of alloying and pick-up between the electrode material and coating. However, work by Williams[21] shows that electrode life is affected very considerably by the type of coating and this is illustrated in Table 2.12. Thus for both the hot dip and electro-deposited coatings, the longest lives are obtained with the iron–zinc and zinc–nickel alloy coatings, values in excess of 5000 welds being typical. This table also shows that the electrode life decreases as the coating thickness increases in the electro-galvanized material. As stated earlier, electrode life is insensitive to small changes in coating thickness in the hot dip product.

Strip steels in automotive applications

The automotive industry is the most important single market for steel strip and has provided the greatest stimulus and challenge for the development of improved or new grades. In spite of competition from aluminium and plastics, steel has maintained its position as the predominant material for automobile body and structural components due to its good formability, ease of welding and relatively low cost. However, during the 1970s and 1980s, significant changes took place in the selection of strip steels for automotive construction due to:

1. The need to reduce the weight of vehicles in order to improve fuel economy.
2. The need to improve the corrosion performance and provide customers with better warranties against structural and cosmetic deterioration.

Table 2.12 *Typical values of electrode life obtained for various 1.0 mm thick zinc coated steels: straight truncated cone electrodes (After Williams[21])*

Coating Type		Electrode Life (No of Welds)
Hot dip zinc	G275	1000–2000
Iron-zinc alloy	IZ100	6000–8000
Electrolytic zinc	2.5μm	10 000
	5.0μm	9000
	7.5μm	7500
	10μm	5000
Zinc-nickel alloy		10 000
Zincrox		10 000

Weight reduction

In the early 1970s, higher strength steels were introduced in the United States for safety-related or structural members such as bumper reinforcement, side door beams and seat belt anchors. These components were manufactured principally from hot-rolled, Nb-treated, micro-alloy steels which provided a favourable cost/weight ratio compared with conventional plain carbon steels. In addition, they necessitated only minor changes in manufacturing methods and facilities. However, the oil crises of 1973 and 1979 provided far greater impetus for the use of higher strength steels, particularly in relation to cold-reduced strip for both inner and outer body panels. Whereas weight saving for improved fuel economy could be achieved by the substitution of conventional steel by plastics and light metals, this will generally lead to an increase in cost, a step taken only reluctantly by the automobile industry. According to Magee,[22] high-strength steels are the only materials that offer the potential for both weight and cost savings, both parameters increasing as the strength of the steel increased. This aspect is illustrated in Figure 2.23.

Whereas the precipitation-strengthened grades were satisfactory for the relatively simple parts mentioned earlier, factors such as reduced formability and increased springback compared with plain carbon steels presented major problems in the introduction of these steels for more complex or shape-sensitive components. In addition, the formability of a material also decreases with reduction in thickness, as indicated by the decrease in the plane strain intercept in the forming limit diagram. Therefore down-gauging only serves to exacerbate the problem of the loss of formability with increased strength.

According to Takechi,[15] precipitation-strengthened or dual-phase, cold-reduced steels cannot be used for outer panels if their $\bar{r}$ value is in the range 1.0–1.3. Rephosphorized steels have a maximum $\bar{r}$ value of about 1.6 and are satisfactory for components other than fenders and deep-drawn parts, provided their yield strength is controlled so as to prevent excessive springback. As indicated earlier, springback increases with yield strength and therefore the

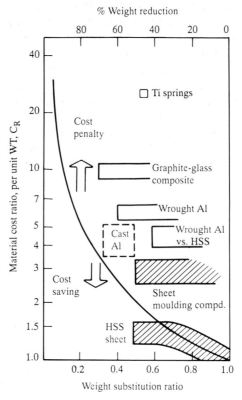

Figure 2.23   *Materials cost ratio (logarithmic scale) against substitution ratio for various materials with mild steel as the base except for the case of wrought Al versus HSS (After Magee[22])*

bake-hardening grades, with their low initial yield strength, represent a major attraction to automotive manufacturers.

   With the introduction of higher strength steels in reduced thicknesses for outer panels, particular attention must be given to dent resistance. Drewes and Engl[23] report on the use of the following equation in order to calculate the increase in yield stress required to compensate for a reduction in thickness and maintain the same plastic dent resistance:

$$\frac{R_{p0.2(1)}}{R_{p0.2(2)}} = \left[\frac{t_{(2)}}{t_{(1)}} \right]^{x}$$

These authors have also presented a relationship between yield strength, sheet thickness and the load to produce a residual dent depth of 0.2 mm. This is reproduced as Figure 2.24. It is reported that rephosphorized, micro-alloy and bake-hardening grades produce a similar effect at a given sheet thickness but dual-phase steels behave differently due to their high rate of work hardening.

Precipitation-strengthened, hot-rolled grades have also been used very successfully for weight reduction in automotive construction, typical applications being chassis members, seat sliders and mountings, rear axle casings, axle tubes, clutch covers and engine mounting brackets. These steels have also been used for many years for automotive wheels but dual-phase steels, with a tensile strength of 590 N/mm², are also used in Japan for these components.[15]

Higher strength steels now account for as much as 40% of the *body-in-white* (structural shell/skin) of vehicles in Japan. On the other hand, the use of these steels has not developed to the same extent in Europe and this can be attributed,

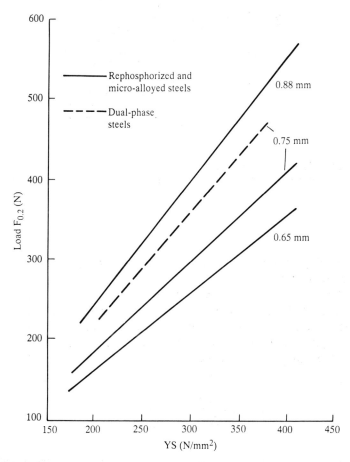

Figure 2.24 *Effect of yield strength and sheet thickness on dent resistance under quasistatic localized loading (F₀.₂ = load producing 0.2 mm remaining dent depth) (After Drewes and Engl[23])*

in part, to the absence of legislation that places an onus on manufacturers to improve the fuel economy of their vehicles.

Improved corrosion resistance

In recent years, the use of zinc-coated steels in car body construction has increased dramatically as manufacturers have attempted to improve their warranties against corrosion. Two types of corrosion have to be considered, namely:

1. Perforation corrosion which takes place from the inside of the vehicle to the outside, causing a hole to appear.
2. Cosmetic corrosion which takes place on the external surfaces of the vehicle.

Most manufacturers have incorporated about 70% coated steel in the body-in-white and it is reported that Audi has adopted a 100% coated steel construction. Both hot dip and electro-galvanized steels are being used in place of uncoated, cold-reduced steel and their application to different parts of the car body has been described by Dasarathy and Goodwin.[13] This is illustrated in Figure 2.25 which refers to the following types of material:

- CR1FF: uncoated steel
- Galvatite: hot dip zinc
- Galvatite IZ: hot dip iron–zinc alloy
- Nizec: electro-deposited 87% Zn 13Ni
- Durasteel: Nizec + 10 µm organic coating

Durasteel is one of a number of proprietary steels with a thin organic coating which is intended primarily as a lubricant for complex forming operations but which also provides corrosion resistance. After forming, the organic coating is on the inside surface of the car where it is subsequently electroprimed but not finished with a surfacer or topcoat.

The substitution of uncoated steel by coated products in automobiles introduces a number of production implications and the experience in the Rover

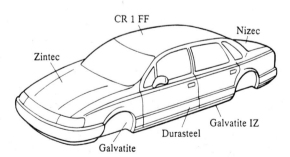

Figure 2.25   *Typical automotive body panels fabricated from metallic coated steels (After Dasarathy and Goodwin[13])*

Group has been described by Thompson.[24] One very important consideration is that the zinc surface is significantly softer than an uncoated steel surface and is therefore more susceptible to damage during decoiling and handling. However, these problems have been overcome at Rover by training programmes rather than facility changes. The soft nature of the zinc coating can also cause zinc to be wiped off the steel by the tools during deep-drawing operations. However, it is reported that the wiping action has little effect on the corrosion behaviour as only a small amount of zinc is actually removed and the coating maintains sacrificial protection. On the other hand, the pick-up of zinc on the tool necessitates more frequent tool cleaning and a reduction in the output of the press.

Thompson reports that the blanking process can leave small particles of zinc around cropped edges which could deposit on the tools and produce surface imperfections during pressing. However, this problem has been overcome by the installation of blank-washing equipment. Although there was concern initially that the introduction of coated steels might require major press tool modifications because of the different frictional characteristics of the surface, no tool modifications were necessary and the problems have been overcome by improved housekeeping and routine maintenance. In relation to the welding implications, the author comments on the need for changes in machine parameters to cope with the increased current (10–20%) requirements of coated steel and also on the greater frequency of weld tip dressing due to the build-up of zinc on the tip surface.

Reference was made earlier to *Durasteel*, which has a 1 μm organic coating over a zinc-rich substrate. However, duplex coatings with organic components of 5–7 μm thick are also finding application in car body construction in order to improve the corrosion performance and resistance to stone-chip damage. Bowen[25] has stated that the incidence of stone-chip damage has increased due to the greater use of road salt and also because of changes in vehicle design. The latter is related to the practice of body streamlining in order to reduce air resistance and the drag coefficient. Thus the paintwork on a sloping bonnet has become subjected to greater attack than was the case on unpainted radiator grilles. This problem is also exacerbated to some extent by the fact that zinc coatings are inferior substrates to bare steel from the point of view of achieving good paint adhesion and therefore stone-chip damage is more evident. Bowen states that this problem can be overcome by using duplex coatings and, as an example, describes the use of *Zincal Duplex*, an electro-galvanized steel coated with *Bonnazine 2000*, an epoxy-based primer, filled with metallic zinc and aluminium and containing molybdenum sulphide as a slip agent during forming. It is reported that Ford are now applying duplex coatings to the bonnets of all Granadas/Scorpios and Fiestas, achieving a 20% reduction in stone-chip damage and the complete elimination of red-rust corrosion.

Dasarathy and Goodwin[13] comment on the need for preprimed steel for automotive construction which would replace the electropriming and surfacer primer stages of finishing at the manufacturers. Such a product would be hot dip or electro-deposited zinc-coated steel with a 25 μm organic coating. However, the authors state that this type of material is still very much at the development stage and one of the major problems that must be overcome is that current organic coatings are not weldable.

Before leaving automotive applications, reference should also be made to the use of terne-coated steel products, i.e. with tin–lead coatings. *Ternex*[26] has a hot dip coating of 92% Pb 8% Sn, applied over a nickel flash, electrolytically deposited over the steel surface. Such materials are resistant to attack by automotive fuels and are used extensively for the manufacture of petrol tanks.

Strip steels in buildings

The construction industry is a major consumer of strip steel but still represents a major growth sector for this product.

Steel-framed houses

Following earlier problems, interest has now been revived in steel-framed housing since it offers many advantages over traditional construction. In particular, the frame can be erected very quickly and made weatherproof so as to allow internal work to proceed at an early date. Labour costs and construction times are also reduced which leads to lower interest costs and faster financial returns. Benefits also accrue to the householder since the steel frame does not absorb moisture and is not subject to shrinkage or warping. The risk of cracking in wall linings is therefore reduced and the steel frame is also fire resistant.

Earlier attempts to introduce steel-framed houses were unsuccessful because the designs were geared to rapid construction rather than aesthetic appeal. The initial steel frames were also manufactured in painted mild steels which were prone to corrosion problems.[27] In current construction, these problems have been overcome by:

1. Preserving a traditional exterior of bricks and mortar around a load-bearing steel frame.
2. Employing hot dip galvanized steel in place of painted steel for corrosion resistance.

An illustration of the construction of a steel-framed house with the preservation of a traditional exterior is shown in Figures 2.26(a) and (b).

The structural members that make up the steel frame are roll-formed U channels in Z28 (280 N/mm² min. YS) grade steel with a zinc coating of 275 g/m² (including both sides). According to Haberfield,[27] modular frames, typically 5 m × 2.4 m high, are fabricated at the factory and bolted together on site on traditional foundations. The internal partitioning walls are constructed and assembled in the same way, followed by the second storey and roof. Steel joists are used to support the floor loadings from the second floor.

The above author states that up to 2 tonnes of hot dip galvanized steel sections may be used in a three-bedroomed house and the thickness of the

Figure 2.26 *Steel-framed housing (Courtesy of Precision Metal Forming Ltd)*

sections ranges from under 1 mm to 2.5 mm or greater, depending on the function. The basic frame is designed to withstand all forces and no strengthening or stiffening contribution is assumed from the external brick or internal plasterboard linings.

Steel cladding

From the early days of 'corrugated iron', strip products have featured prominently as cladding materials, particularly in the construction of industrial buildings. A considerable amount of effort has been devoted to the development of structurally efficient profiles for steel cladding and some typical profiles are shown in Figure 2.27.

Although a significant part of the market for cladding and roofing is still satisfied by unpainted, or zinc- or zinc–aluminium-coated strip, this area has been revolutionized by the introduction of pre-finished, organic-coated strip. These products were described in an earlier section and involve a variety of paint or plastic formulations which are applied to zinc-coated steels. These coatings provide enhanced corrosion resistance and are available in a wide range of colours and surface textures.

The types of coating that are recommended for external use include PVC Plastisol (200 μm), PVF 2 (27 μm), Silicone Polyester (25 μm) and Architectural Polyester (25 μm).[16,17]

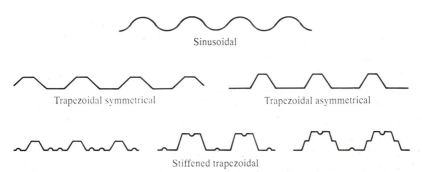

Figure 2.27 *Typical profiles of organic-coated steel cladding (After Lewis et al.[28])*

Steel lintels

Lightweight steel lintels have now virtually replaced reinforced concrete lintels in domestic housing. These components are specified in BS 5977: Part 2: 1983 *Specification for prefabricated lintels*, which provides guidance on the design and testing of all types of lintel for use in domestic buildings up to three storeys high. This specification deals with post-galvanized lintels with heavy zinc coatings and also those fabricated from pre-galvanized strip with minimum zinc coatings of up to 600 g/m² (including both sides).

Other applications for strip steels

Domestic appliances

On page 42, information was provided on the production and properties of pre-finished strip, coated with various types of paint or plastic. During the late 1960s, work was initiated on the use of pre-painted strip for the high-gloss, white domestic appliance market and this proved to be a major commercial success. Thus the majority of wrap-arounds for washing machines, refrigerators, dish washers, domestic boilers and freezers are now produced from pre-painted strip, providing the following benefits to the manufacturers:

1. Higher productivity.
2. Savings in floor space.
3. Reduced capital expenditure.
4. Elimination of the effluent problems associated with paint work.

The organic coating used for this type of work is a high-gloss polyester, 25 µm thick over a galvanized steel base, which provides a high flexibility/hardness ratio and good resistance to elevated temperatures and detergents. Although the degree of cold forming is not very severe for the majority of the components involved, areas such as the ports for front-loading washing machines are subjected to a limited drawing operation.

Steel drums

Steel drums, with capacities ranging from 25 to 210 litres, constitute a major market for cold-reduced strip but one that is now facing major competition from plastics. The formability requirements are not high and steel drums are normally made from uncoated, CR4 strip. However, limited use is made of galvanized steel to provide greater corrosion resistance against particular products.

For the future, there is the prospect of using higher strength steels in lighter gauges in an attempt to stem the greater penetration of the market by plastics. However, attention must also be given to improved designs in order to preserve adequate rigidity in down-gauged drums.

Vitreous enamelled products

Vitreous enamelling is a long-established finishing process and one that is being applied to an increasing range of steel products. Cold-reduced steel is the most widely used substrate for enamelling, ranging from CR1 to CR4 according to the end application. After fabrication, the product is degreased, pickled, rinsed and dried and a ground coat enamel slip or slurry is applied to the component, either by dipping or spraying. After drying, the coating is fused to the steel by firing at temperatures above 800°C. A second coat, the cover coat enamel, is then applied, dried and fired in a similar manner.

The process provides a hard, durable finish, in a range of attractive colours, which is resistant to heat, abrasion and chemical attack. The applications for vitreous enamelled steel include domestic cookers, baths, hot-water tanks and cookware, architectural facing panels, flue pipes plus silos and tanks for bulk storage.

References

1. Davies, G.M. and Thompson, S.J.A. *Steel Times*, April, 180 (1988).
2. Lankford, W.T., Snyder, S.D. and Bauscher, J.A. *Trans. ASM*, **42**, 1197 (1950).
3. Backhofen, S.P. and Keeler, W.A. *Trans. ASM*, **56**, 25 (1963).
4. Miles, J. *Metals and Materials*, July, 398 (1989).
5. Hudd, R.C. In *Proc. Low Carbon Structural Steels for the Eighties*, Spring Residential Course Series 3 No. 6, The Institution of Metallurgists, March 1977, 111A–1.
6. Hudd, R.C. *Metals and Materials*, February, 71 (1987).
7. *Steelresearch 76*, British Steel, 7 (1976).
8. Ono, S., Nozoe, O., Shimomura, T. and Matsudo, K. In *Proc. Metallurgy of Continuous Annealed Sheet Steel* (eds Bramfitt, B.L. and Manganon, P.L) AIME, Dallas, p. 99 (1982).
9. Toda, K., Gondoh, H., Takechi, H., Abe, M., Uehora, N. and Kominya, T. *Tetsu-to-Hagane*, **61**, 2363 (1975).
10. Takahashi, N., Shibata, M., Furuno, Y., Hayakawa, H., Kakuta, K. and Yamamoto, K. In *Proc. Metallurgy of Continuous Annealed Sheet Steel* (eds Bramfitt, B.L. and Manganon, P.L.), AIME, Dallas, p. 133 (1982).
11. Committee on Technology *High Strength Low Alloy Steels*, International Iron and Steel Institute, Brussels (1987).
12. Bordignon, P.J.P., Hulka, K. and Jones, B.L. *High Strength Steels for Automotive Applications*, Niobium Technical Report NbTR-06/84, December (1984).
13. Dasarathy, C. and Goodwin, T.J. *Metals and Materials*, January, 21 (1990).
14. Rashid, M.S. *SAE Tech. Paper 770211* (1977).
15. Takechi, H. In *Proc. Vehicle Design and Components – Materials Innovation and Automotive Technology*, 5th IAVD Conference (ed. Dorgham, M.A.), Geneva, Interscience (1990).

16. British Steel *Colourcoat and Stelvetite Brochure*, British Steel.
17. British Steel *Colourcoat and Stelvetite Pre-finished Steel Technical Manual*, British Steel.
18. Morgan, E. *Tinplate and Modern Canmaking Technology*, Pergamon Press (1985).
19. Jones, T.B. and Williams, N.T. *Resistance Spot Welding of High Strength Steels*, C287/81, I. Mech. E. (1981).
20. Jones, T.B. and Williams, N.T. Resistance spot welding of rephosphorised steels, *A Review of Welding in the World*, **23**, No. 11/12, 248.
21. Williams, N.T. In *Proc. Vehicle Design and Components – Materials Innovation and Automotive Technology*, 5th IAVD Conference (ed. Dorgham, M.A.), Geneva, Interscience (1990).
22. Magee, C.L. *SAE Preprint 820147* (1982).
23. Drewes, E.J. and Engl, B. In *Proc. Coated and High Strength Steels – AutoTech 89*, I. Mech. E., C399/37 (1989).
24. Thompson, S.J.A. In *Proc. Coated and High Strength Steels – AutoTech 89*, I. Mech. E., C399/37 (1989).
25. Bowen, M. In *Proc. Coated and High Strength Steels – AutoTech 89*, I. Mech. E., C399/37 (1989).
26. British Steel *Ternex brochure*, British Steel.
27. Haberfield, A.B. *Steelresearch 1987–88*, British Steel, 42 (1988).
28. Lewis, K.G., Jones, D. and Godwin, M.J. *Metals and Materials*, June 357 (1988).

3 Low-carbon structural steels

Overview

Although open to wide interpretation, the term *structural steels* is commonly used to identify the predominantly C–Mn steels, with ferrite–pearlite micro-structures, which are used in large quantities in civil and chemical engineering. The steels are produced in plates and sections, sometimes up to several inches thick, and generally with yield strength values up to about 500 N/mm^2. However, structural steels also include low-alloy grades which are quenched and tempered in order to provide yield strengths up to about 700 N/mm^2.

As illustrated in this chapter, these steels are used in a wide and diverse range of applications, including buildings, bridges, pressure vessels, ships and off-highway vehicles. More recently, structural steels have been used extensively in very demanding applications such as offshore oil and gas platforms and the associated pipelines that often operate in extremely cold and chemically aggressive environments.

A major feature of most forms of construction in structural steels is the high level of welding employed and the requirement for high-integrity welds. Welding began to replace riveting as the principal joining process in the 1940s but, at that time, structural steels were characterized by high carbon contents and were therefore prone to cold cracking. The requirement for lower carbon grades with improved weldability was illustrated very dramatically in the construction of the first all-welded merchant ships (*Liberty ships*) during World War II. However, the break-up of these vessels on the high seas also led to the recognition of a further major property requirement in structural steels, namely toughness as opposed to ductility.

In the early 1950s, the work of Hall and Petch revolutionized the design of structural steels with the concept that refinement of the ferrite grains led to an increase in both the yield strength and toughness of ferrite–pearlite steels. Thus steels with yield strength values up to about 300 N/mm^2 could be produced in aluminium-grain-refined compositions, with good impact properties and with good welding characteristics. Ferrite grain refinement remains the single most important metallurgical parameter in the make-up of modern structural grades but the demand for higher strength steels required a further strengthening mechanism, namely precipitation strengthening. Thus small additions of nio-bium, vanadium and titanium were added to structural steels to raise the yield strength up to a level of about 500 N/mm^2. Since they were added in levels of up to only 0.15%, these additions became known as *micro-alloying* elements and the compositions were designated *High-strength Low-alloy* (HSLA) steels.

The late 1950s and 1960s represented a period of major research on the structure–property relationships and fracture behaviour of structural steels. However, it also heralded the introduction of an important new technique in the production of structural steels, namely *controlled rolling*. In essence, this enabled fine-grained steels to be produced in the as-rolled condition, thereby eliminating the need for costly normalizing heat treatments. More importantly, controlled rolling led to the generation of steels with properties far superior to those that could be obtained in the normalized condition.

In the 1970s and 1980s, controlled rolling was augmented with controlled cooling and the combination is now referred to as *thermomechanical processing*. In its more severe form (*direct quenching*), controlled cooling is now used as an alternative to reheat quenching for the production of quenched and tempered grades.

Structural steels have therefore undergone very significant changes, each change producing a substantial improvement in an important property such as strength, toughness or weldability. Aspects such as improved cleanness and inclusion shape control have also been adopted, leading to improvements in fabrication and service performance. These factors, coupled with very favourable cost comparisons, have meant that structural steels have remained virtually unchallenged by competitive materials, other than reinforced concrete, in most of their traditional applications.

Strengthening mechanisms in structural steels

Major research effort has been devoted to the detailed understanding of factors affecting the properties of low-carbon structural steels. Whereas considerable cost savings accrued from the use of lighter sections in higher strength steels, there was also the need to maintain, or indeed improve upon, other important properties such as toughness and weldability. Therefore detailed attention was given to identifying the strengthening mechanisms which were most cost-effective or that provided the best combination of properties.

The practical options for increasing the strength of steels are:

1. Refining the ferrite grain size.
2. Solid solution strengthening.
3. Precipitation strengthening.
4. Transformation strengthening.
5. Dislocation strengthening.

Whereas work hardening or dislocation strengthening can result in very high levels of strength, these are achieved at the expense of toughness and ductility. For this reason, little use is made of this method of strengthening but, as indicated in Chapter 2, work hardening finds limited application in low-carbon strip grades in the form of recovery-annealed steels. As illustrated later in this chapter, work hardening is also used in the production of high-strength reinforcing bars.

Ferrite grain refinement

In the early 1950s, work published by Hall[1] and Petch[2] laid the foundation for the development of modern, high-strength structural steels. The Hall–Petch equation, perhaps the most celebrated in ferrous metallurgy, is as follows:

$$\sigma_y = \sigma_i + k_y d^{-\frac{1}{2}}$$

where

σ_y = yield strength
σ_i = friction stress which opposes dislocation movement
k_y = a constant (often called the dislocation locking term)
d = ferrite grain size

Thus refinement of the ferrite grain size will result in an increase in yield strength and the relationship is shown in Figure 3.1.

Whereas a strengthening effect usually leads to a decrease in toughness, it was shown that refinement of the ferrite grain size also produced a simultaneous improvement in toughness. The Petch equation linking toughness to grain size is given below:

$$\beta T = \ln \beta - \ln C - \ln d^{-\frac{1}{2}}$$

where β and C are constants, T is the ductile–brittle transition temperature and d is the ferrite grain size. Therefore, as illustrated in Figure 3.1, the impact transition temperature decreases as the ferrite grain size is reduced.

Refinement of the ferrite grain size can be achieved in a number of ways. Traditionally, fine-grained steels contain about 0.03% Al which is soluble at

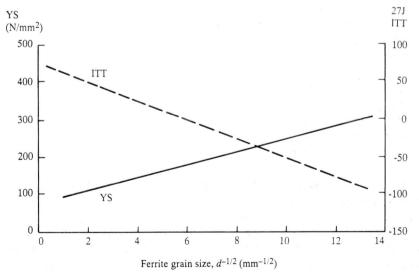

Figure 3.1 *Effect of ferrite grain size on yield strength and impact properties*

normal slab or bloom reheating temperatures of around 1250°C and which remains in solution during rolling and after cooling to ambient temperature. However, on subsequent reheating through the ferrite range to the normalizing or solution treatment temperature, the aluminium combines with nitrogen in the steel to form a fine dispersion of AlN. These particles pin the austenite grain boundaries at the normal heat treatment temperatures just above Ac_3 (typically 850–920°C, depending upon carbon content) and therefore result in the formation of a fine austenite grain size. In turn, a fine austenite grain size results in the formation of a fine ferrite grain size on cooling to room temperature.

Although the austenite grain size is of major importance, other factors also play a part in developing a fine ferrite grain size. Thus the addition of elements such as carbon and manganese or an increase in the cooling rate from the austenite temperature range will lead to a refinement of the ferrite grains. In either case, this is achieved by depressing the temperature of transformation of austenite to ferrite. However, there is obviously a limit to the amount of strengthening that can be obtained by this mechanism before transformation is depressed to such an extent that it leads to the formation of bainite or martensite and the introduction of transformation strengthening.

Solid solution strengthening

The solid solution strengthening effects of the common alloying elements are illustrated in Figure 3.2 and work by Pickering and Gladman[4] has provided the strengthening coefficients shown in Table 3.1 for ferrite–pearlite steels containing up to 0.25% C and 1.5% Mn. These data illustrate the very powerful strengthening effects of the interstitial elements, carbon and nitrogen, but it must be borne in mind that these elements have only a very limited solid solubility in ferrite. However, both carbon and nitrogen also have a very adverse effect on toughness. Of the substitutional elements, phosphorus is the most potent and, as indicated in Chapter 2, additions of up to about 0.1% P are

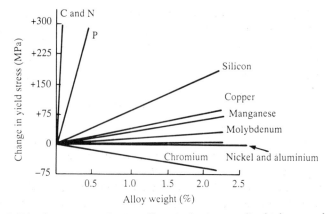

Figure 3.2   *Solid solution strengthening effects in ferrite–pearlite high-strength low-alloy steels (After Pickering[3])*

Table 3.1

Element	N/mm^2 per 1 Wt %
C and N	5544
P	678
Si	83
Cu	39
Mn	32
Mo	11
Ni	~ 0
Cr	-31

incorporated in the higher strength rephosphorized grades that are used in automotive body panels. However, like carbon and nitrogen, phosphorus has a detrimental effect on toughness and therefore it is not used as a strengthening agent *per se* in structural steels. On the other hand, phosphorus is added to the so-called *weathering grades* because of its beneficial effect on atmospheric corrosion resistance. These steels will be discussed later in this chapter. Of the remaining elements, only silicon and manganese are cost-effective as solid solution strengtheners but silicon is added to steels primarily as a deoxidizing agent.

Precipitation strengthening

Precipitation strengthening can be induced by a variety of elements but, in the context of ferrite–pearlite steels, the systems of commercial significance are those involving niobium, vanadium and titanium. These elements have a strong affinity for carbon and nitrogen and, consequently, they have only a limited solid solubility in steel. Therefore they are added to steels in small amounts, e.g. up to about 0.06% Nb or 0.15% V, and as stated earlier are often referred to as micro-alloying elements.

As illustrated in Figure 3.3(a), a substantial amount of niobium will be taken into solution at a slab or bloom reheating temperature of 1250°C. On cooling, Nb(CN) will precipitate at the austenite–ferrite interface during transformation (*interphase precipitation*) which leads to substantial strengthening. On the other hand, on reheating to a typical normalizing temperature of 920°C, very little Nb(CN) will dissolve and therefore virtually no precipitation strengthening can take place. However, the undissolved particles will act as pinning agents, restricting austenite grain growth and leading to the formation of a fine ferrite grain size. Therefore the reheating temperature controls the potential for precipitation strengthening and the strength increases progressively as the temperature is raised from 920 to 1250°C.

As indicated in Figure 3.3(b), vanadium dissolves more easily than niobium and complete solution of V_4C_3 would be expected to occur in commercial grades of structural steel at typical normalizing temperatures, e.g. 920°C. Slightly higher temperatures are required for the solution of VN which can act as a

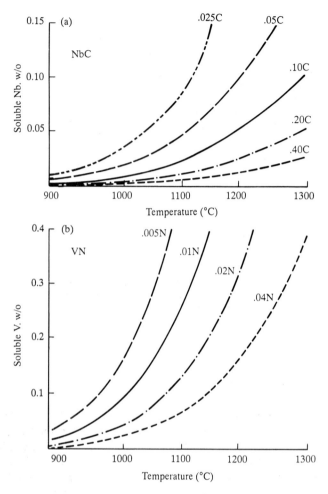

Figure 3.3 *Solubility of NbC and VN in austenite at various temperatures (After Irvine et al.[5])*

grain-refining agent at a temperature of 920°C. However, in Al–V steels, aluminium is the more powerful nitride former, and in the presence of 0.04% Al, significant levels of vanadium will go into solution at 920°C and be available for the precipitation of V_4C_3 on transformation to ferrite. Vanadium steels therefore provide significant precipitation-strengthening effects, i.e. up to 150 N/mm² per 0.10% V.

The strengthening effect of precipitated particles is dependent on both the volume fraction and particle size of the precipitates. This is illustrated in Figure 3.4, which was derived by Gladman *et al.* using the Ashby–Orowan model for precipitation strengthening. Whereas the volume fraction of precipitate is controlled by aspects such as solute concentration and solution treatment temperature, the particle size will be influenced primarily by the temperature of transformation, which is controlled by the alloy content and cooling rate effects.

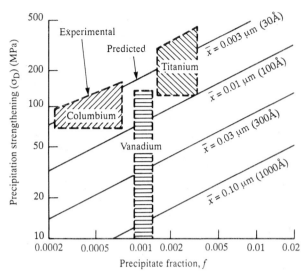

Figure 3.4   *The dependence of precipitation strengthening on precipitate size (x̄) and fraction according to the Ashby–Orowan Model, compared with experimental observations for given micro-alloying additions (After Gladman* et al.[6])

Transformation strengthening

As stated earlier, both alloying elements and faster cooling rates depress the temperature of transformation of austenite to ferrite and, ultimately, the effect will be sufficient to cause transformation to bainite or martensite. The consequence of this progression is illustrated in Figure 3.5, which relates to steels containing 0.05–0.20% C. Thus the strength is increased progressively with the introduction of lower temperature transformation products but, of course, with some sacrifice to toughness and ductility. However, in the context of structural steels, there is a demand for quenched and tempered low-alloy grades with yield strengths up to 700 N/mm². Such steels are normally alloyed with molybdenum and boron to promote hardenability but there may also be a need to include elements such as vanadium to improve tempering resistance.

Structure–property relationships in ferrite–pearlite steels

Following the derivation of the Hall–Petch relationship:

$$\sigma_y = \sigma_i + k_y d^{-\frac{1}{2}}$$

it was proposed that this basic equation for yield strength could be extended to take account of the strengthening effects of alloying elements. Thus:

$$\sigma_y = \sigma_i + k'(\% \text{ alloy}) + k_y d^{-\frac{1}{2}}$$

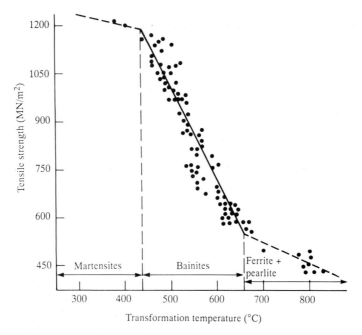

Figure 3.5 *Relationship between 50% transformation temperature and tensile strength (After Pickering[3])*

During the late 1950s and early 1960s, Gladman and Pickering[4] pursued this line of development and provided the following quantitative relationships for yield strength, tensile strength and impact transition temperature:

$$\text{YS (N/mm}^2) = 53.9 + 32.3\% \text{ Mn} + 83.2\% \text{ Si} + 354\% \text{ N}_f + 17.4d^{-\frac{1}{2}}$$

$$\text{TS (N/mm}^2) = 294 + 27.7\% \text{ Mn} + 83.2\% \text{ Si} + 3.85\% \text{ pearlite} + 7.7d^{-\frac{1}{2}}$$

$$\text{ITT (}^\circ\text{C)} = -19 + 44\% \text{ Si} + 700 \sqrt{(\%\text{N}_f)} + 2.2\% \text{ pearlite} - 11.5d^{-\frac{1}{2}}$$

where d is the mean ferrite grain size in mm and N_f the free (soluble) nitrogen.

Each of these equations illustrates very clearly the beneficial effects of a fine ferrite grain size in increasing the yield and tensile strength and depressing the impact transition temperature. It is also interesting to note that the pearlite content has no significant effect on the yield strength of these low-carbon, predominantly ferritic steels. On the other hand, pearlite increases the tensile strength and has a detrimental effect on toughness. The solid solution strengthening effects of manganese, silicon and free nitrogen are also highlighted in the above equations and it will be noted that free nitrogen is particularly detrimental to the impact properties.

Whereas the above equations clearly identify the solid solution strengthening effects of an element such as manganese, it must be borne in mind that manganese also contributes to strength by other means. Thus by depressing the temperature of transformation of austenite to ferrite, manganese causes further strengthening by:

1. Refining the ferrite grain size.
2. Refining the size of precipitation-strengthening particles, e.g. Nb(CN) and V_4C_3.

These effects are illustrated in Figure 3.6, which shows the effect of manganese on the yield strength of a V–N steel, normalized from 900°C. This figure also indicates that free nitrogen contributes very little to the overall strength of this

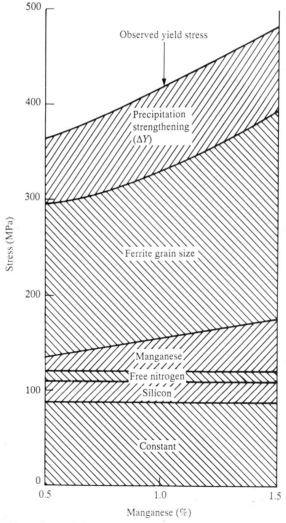

Figure 3.6 *The effect of increasing manganese content on the components of the yield stress of steels containing 0.2% carbon, 0.2% silicon, 0.15% vanadium and 0.15% nitrogen, normalized from 900°C (After Gladman et al.[6])*

particular steel in spite of having a very large strengthening coefficient. This is due to the fact that most of the nitrogen in this particular alloy remains out of solution as VN at a temperature of 900°C and therefore very little free nitrogen is available for solid solution strengthening. Whereas the VN particles refine the austenite grain size and produce a fine ferrite grain size, vanadium in solution leads subsequently to a dispersion-strengthening effect of the order of 75 N/mm^2.

Controlled rolling/thermomechanical processing

As stated earlier, the traditional route to a fine grain size in ferrite–pearlite structural steels has been to incorporate grain-refining elements, such as aluminium, and to normalize the materials from about 920°C after rolling. However, prior to the introduction of continuous casting, basic carbon steel plate was made from semi-killed (*balanced*) ingots and the additional costs associated with aluminium grain refinement were very considerable, as illustrated by the cost data from the early 1960s shown in Table 3.2.

Table 3.2

Requirement	Extras (£/tonne)	Accumulative price Price (£/tonne)
nil (mild steel)	–	42.62
Si killing	7.50	50.12
Grain refinement	3.00	53.12
Normalizing	3.00	56.12
Impact testing <0°C	3.50	59.62

In the late 1950s, steel users were also gaining experience with Nb-treated, micro-alloy steels which provided substantially higher strengths than plain carbon steel in the as-rolled condition but with a significant reduction in toughness compared with aluminium-grain-refined steels. However, when the micro-alloy steels were normalized to improve their impact properties, their strength advantage was forfeited. There was therefore the need for an alternative route to a fine grain size in structural steel plate which would overcome both the cost and strength penalties associated with traditional normalizing. In fact, the first indication of a viable alternative to normalizing was published in 1958 when Vanderbeck[7] reported that European steel producers were adopting lower than normal finishing temperatures during rolling, in order to refine the structure and improve mechanical properties. This practice became known as *controlled rolling* but in more recent years, the term *thermomechanical processing* has been used increasingly to embrace both modified hot-rolling and in-line accelerated-cooling operations.

Outline of process

The traditional hot-rolling operation for plates is shown schematically in Figure 3.7(a). Typically, slabs are soaked at temperatures of about 1200–1250°C and these are rolled progressively to lower plate thicknesses, often finishing at temperatures above 1000°C. In plain carbon steels, soaking at 1200–1250°C

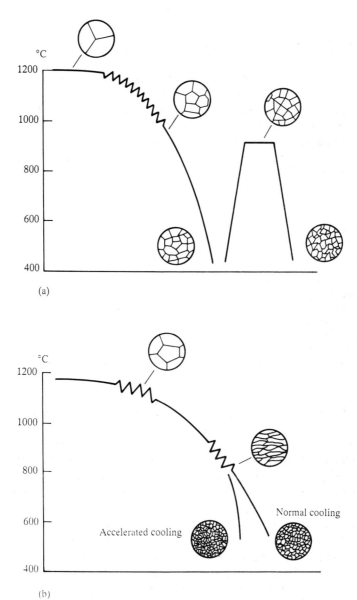

(a)

(b)

Figure 3.7   *Rolling schedules; (a) normal rolling and normalizing; (b) thermomechanical processing*

produces a coarse austenite grain size and rolling just below that range results in rapid recrystallization. Even at a finishing temperature of 1000°C, recrystallization and subsequent grain growth will be relatively rapid, resulting in the generation of a coarse austenite grain size. On cooling to room temperature, this results in the formation of a coarse ferrite grain size and the material must be normalized in order to refine the microstructure.

In controlled rolling, the operation is a two-stage process and, as illustrated in Figure 3.7(b), a time delay is introduced between roughing and finishing. This allows the finishing operations to be carried out at temperatures below the recrystallization temperature, which results in the formation of fine *pancaked* austenite grains and transformation to a fine-grained ferrite structure. In the 30 or so years since its introduction, a considerable amount of research work has been carried out world-wide on controlled rolling which has led to the development of materials with vastly improved mechanical properties compared with those obtained in conventional, heat-treated steels. Whereas the early experiments in controlled rolling were carried out on plain carbon steels, it soon became evident that the process was greatly facilitated by the addition of carbide-forming elements. In particular, it was shown that the addition of about 0.05% Nb caused a marked retardation in recrystallization which allowed controlled rolling to be carried out at significantly higher temperatures.

Slab reheating

The slab-reheating stage is important in that it controls the amount of micro-alloying elements taken into solution and also the starting grain size. The solubility curves for NbC and VN in steels of different carbon and nitrogen contents are shown in Figure 3.3 and were referred to earlier. For most commercial grades of steel, Sellars[8] states that complete solution of VC is expected at the standard normalizing temperature of 920°C, and VN at somewhat higher temperatures, whereas Nb(CN), AlN and TiC require temperatures in the range 1150–1300°C. TiN is the most stable compound and little dissolution is expected to take place at normal reheating temperatures. Whereas the presence of fine, undissolved carbonitride particles will serve to maintain a fine austenite grain size at the reheating stage, it is equally important that micro-alloying elements are taken into solution so as to be available for the control of recrystallization and precipitation strengthening at later stages in the process. This dual requirement is often achieved by making multiple micro-alloy additions, incorporating the less soluble elements such as niobium and titanium for grain size control during reheating together with vanadium, which dissolves more readily and which provides substantial precipitation strengthening.

Rolling

Tamura *et al.*[9] recognize three distinct stages during controlled rolling:

1. Deformation in the recrystallization temperature range just below the reheating temperature.

2. Deformation in the temperature range between the recrystallization temperature and Ar_3.
3. Deformation in the two-phase $\gamma + \alpha$ temperature range between Ar_3 and Ar_1.

At temperatures just below the reheating temperature, the rate of recrystallization is rapid, increasing with both temperature and degree of deformation. However, refinement of the austenite structure is produced by successive recrystallization between passes, provided the strain per pass exceeds a minimum critical level. Recrystallization is retarded to some extent by the presence of solute atoms, such as aluminium, niobium, vanadium and titanium, and the process is known as *solute drag*. However, the major effect of niobium and other micro-alloying elements in retarding recrystallization and grain growth arises from the strain-induced precipitation of fine carbonitrides during the rolling process.

As the rolling temperature decreases, recrystallization becomes more difficult and reaches a stage where it effectively ceases. Cuddy[10] has defined the *recrystallization stop temperature* as the temperature at which recrystallization is incomplete after 15 seconds, after a particular rolling sequence. Using this criterion, the effect of the common micro-alloying elements on recrystallization behaviour is shown in Figure 3.8 and this illustrates the very powerful effect of niobium. The retardation effects of the various elements are dependent on their relative solubilities in austenite, the least soluble (niobium) having the largest driving force for precipitation at a given temperature and creating a proportionally greater effect in raising the recrystallization temperature than the more soluble elements such as aluminium and vanadium.

By introducing a delay between roughing and finishing, rolling can be made to take place at a temperature below 950°C, where the strain-induced precipitation of Nb(CN) or TiC is sufficiently rapid to prevent recrystallization before the next pass. According to Cohen and Hansen,[11] austenite recrystallization and carbonitride precipitation are interlinked during this process, substructural

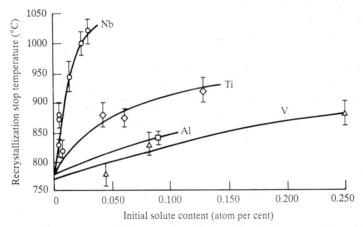

Figure 3.8 *Effect on* recrystallization stop temperature *of increase in micro-alloy content in 0.07% C 1.4% Mn 0.25% Si steel (After Cuddy[10])*

features in the deformed austenite providing nucleation sites for carbonitride precipitation which in turn pins the substructure and inhibits recrystallization. This results in an elongated pancake morphology in the austenite structure and the austenite is said to be *conditioned*. The deformation substructure that is introduced within the austenite grains has a particularly beneficial effect in developing a finer grain size. This arises from the fact that the substructure provides intragranular sites for ferrite nucleation in addition to those at the austenite grain boundaries.

The controlled-rolling operation can be intensified by depressing the deformation process below Ar_3 and into the two-phase $\gamma + \alpha$ region. In addition to further grain refinement, rolling in this region also produces a significant change in the microstructure. Thus a mixed structure is produced, consisting of polygonal ferrite grains which have transformed from deformed austenite and deformed ferrite grains which were produced during the rolling operation.

Transformation to ferrite

Although the mean ferrite grain size is related to the thickness of the pancaked austenite grains, other factors also play an important part in the control of the finished microstructure and properties. As discussed previously, alloying elements depress the austenite to ferrite transformation temperature and thereby decrease the ferrite grain size. A further important effect is the rate of cooling from the austenite (or $\gamma + \alpha$) range and a combination of controlled rolling and accelerated cooling is now being used to produce further improvement in properties.

The benefits of accelerated cooling can be used in two ways:

1. To increase the strength compared with air-cooled, controlled-rolled material.
2. To achieve the strength levels of controlled-rolled materials in steels of lower alloy content.

The latter approach is particularly attractive in that it utilizes steels of lower carbon equivalent and therefore provides improved weldability. This effect is shown schematically in Figure 3.9.

An extension of normal accelerated cooling after rolling is that employing the faster cooling rates of *direct quenching*. Whereas the former is concerned with refinement of the ferrite grains, the latter is concerned with the formation of lower temperature transformation products such as bainite and martensite. Direct quenching avoids the reheating costs associated with conventional off-line hardening treatments but still requires a subsequent tempering treatment.

Standard specifications

For many years, the relevant UK specification for structural steels was BS 4360 *Weldable structural steels*. However, since 1990, parts of BS 4360

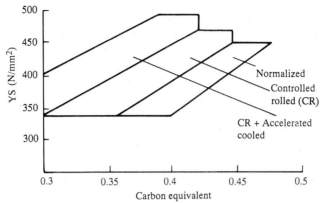

Figure 3.9 *Exploitation of controlled rolling and accelerated cooling for improved properties in plate steels*

have been withdrawn progressively and superseded by the following European specifications:

BS EN 10025: 1993   *Hot rolled products of non-alloy structural steels.*
BS EN 10113: 1993   *Hot rolled products in weldable fine grain structural steels.*

BS 4360: 1990 still contains a residue of grades not yet incorporated in European standards but, eventually, this specification will be withdrawn completely. However, because the move to European standards is incomplete and reference is still made to the last complete version of BS 4360 (1986) in the standards for bridges (BS 5400) and buildings (BS 5950), the author has elected to focus attention initially on this virtually obsolete but still very familiar British standard.

Brief details of the chemical compositions of steels contained in BS 4360: 1986 are shown in Table 3.3. This indicates that there were four basic grades (40, 43, 50 and 55), the numbers referring to the minimum tensile strengths in kgf/mm^2, the stress units in use when the specification was first introduced. As shown in Table 3.3, the lower strength grades were based on C-Mn compositions, whereas Grades 50 and 55 relied on the use of micro-alloying elements to achieve the higher strength values.

Within each strength grade, there were a number of sub-grades which represent increasing levels of impact strength, designated as follows:

Minimum Charpy V value of 27 J at:

A	No requirement	DD	−30°C
B	20°C	E	−40°C
C	0°C	EE	−50°C
D	−20°C	F	−60°C

Thus BS 4360 Grade 50D referred to a steel with a minimum yield strength level of 355 N/mm^2 and a minimum Charpy energy value of 27 J at −20°C.

Table 3.3 *BS 4360: 1986 Weldable structural steels*

Grade	C% max.	Mn% max.	Nb%	V%	TS (N/mm²)	YS min[a] (N/mm²)
40	0.16–0.22	1.5	–	–	340–500	235
43	0.16–0.22	1.5	–	–	430–580	275
50	0.16–0.25	1.5	0.003–0.1	0.003–0.1	490–640	355
55	0.16–0.25	1.5	0.003–0.1	0.003–0.1	550–700	450

[a] For plates up to 16 mm.

Note that BS 4360 is being phased out in favour of European specification BS EN 10025 and that BS 4360: 1986 is now obsolete.

The comprehensive yield strength-toughness matrix provided by BS 4360: 1986 is shown in Figure 3.10.

The designations and property requirements of the new European specifications for structural steels are shown in Table 3.4(a) (BS EN 10025: 1993) and Table 3.4(b) (BS EN 10113: 1993). These tables also provide a useful comparison with the grades specified in the former BS 4360 standard. A notable feature of the new designations is that numerical elements refer to the minimum yield strength values in N/mm² whereas those in BS 4360 defined the minimum tensile strengths in kgf/mm². Given that design in structural steels is normally based on yield rather than tensile strength, this is welcomed change but many will regret the passage of BS 4360.

A new standard has also been introduced for *weathering steels* and this will be discussed later in this chapter.

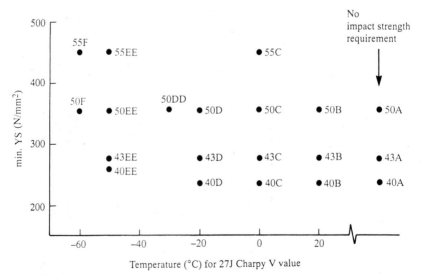

Figure 3.10 *Yield strength–impact strength requirements in BS 4360: 1986 Weldable structural steels (product forms – plates, strip and wide flats)*

Table 3.4(a) *Structural steels: comparison between grades in BS EN 10025: 1993 and BS 4360: 1986*

Grade		Former grade		Tensile strength ≥ 3 mm ≤ 100 mm N/mm²	Min. yield strength at 16 mm N/mm²	Max. Thk for specified yield N/mm²	Impact Energy (J°C) Nominal thickness			Grade	Tensile strength ≤ 100 mm N/mm²	Min. yield strength at 16 mm N/mm²	Max. Thk for specified yield N/mm² (2)	Impact Energy (J°C) Nominal thickness	
							Temp. °C	≤ 150 mm (1)	> 150 mm ≤ 250 mm (1)					Temp. °C	≤ 100 mm (3)
S185	(4)	Fe 310-0	(4)	290/510	185	25	–	–	–	–	–	–	–	–	–
S235	(5)	Fe 360A	(5)	340/470	235	250	+20	–	–	40A	340–500	235	150	–	–
S235JR	(4)	Fe 360B	(4)	340/470	235	25	+20 (6)	27	–	–	–	–	–	–	–
S235JRG1	(4)	Fe 360B(FU)	(4)	340/470	235	25	+20 (6)	27	–	–	–	–	–	–	–
S235JRG2		Fe 360B(FN)		340/470	235	250	+20 (6)	27	23	40B	340/500	235	150	+20 (6)	27
S235JO		Fe 360C		340/470	235	250	0	27	23	40C	340/500	235	150	0	27
S235J2G3		Fe 360D1		340/470	235	250	–20	27	23	40D	340/500	235	150	–20	27
S235J2G4		Fe 360D2		340/470	235	250	–20	27	23	40D	340/500	235	150	–20	27
S275	(5)	Fe 430A	(5)	410/560	275	250	–	–	–	43A	430/580	275	150	–	–
S275JR		Fe 430B		410/560	275	250	+20 (6)	27	–	43B	430/580	275	150	+20 (6)	27
S275JO		Fe 430C		410/560	275	250	0	27	23	43C	430/580	275	150	0	27
S275J2G3		Fe 430D1		410/560	275	250	–20	27	23	43D	430/580	275	150	–20	27
S275J2G4		Fe 430D2		410/560	275	250	–20	27	23	43D	430/580	275	150	–20	27
S355	(5)	Fe 510A	(5)	490/630	355	250	–	–	–	50A	490/640	355	150	–	–
S355JR		Fe 510B		490/630	355	250	+20 (6)	27	–	50B	490/640	355	150	+20 (6)	27
S355JO		Fe 510C		490/630	355	250	0	27	23	50C	490/640	355	150	0	27
S355J2G3		Fe 510D1		490/630	355	250	–20	27	23	50D	490/640	355	150	–20	27
S355J2G4		Fe 510D2		490/630	355	250	–20	27	23	50D	490/640	355	150	–20	27
S355K2G3		Fe 510DD1		490/630	355	250	–20	40	33	50DD	490/640	355	150	–30	27
S355K2G4		Fe 510DD2		490/630	355	250	–20	40	33	50DD	490/640	355	150	–30	27

Table 3.4(a)   *Continued*

		BS EN 10025: 1993						BS 4360: 1986				
Grade	Former grade	Tensile strength ≥ 3 mm ≤ 100 mm N/mm²	Min. yield strength at 16 mm N/mm²	Max. Thk for specified yield N/mm²	Impact Energy (J°C) Nominal thickness Temp. °C	≤ 150 mm (1)	> 150 mm ≤ 250 mm (1)	Grade	Tensile strength ≤ 100 mm N/mm²	Min. yield strength at 16 mm N/mm²	Max. Thk for specified yield N/mm² (2)	Impact Energy (J°C) Nominal thickness Temp. °C ≤ 100 mm (3)
E295	Fe 490-2	470/610	295	250	—	—	—	—	—	—	—	—
E335	Fe 590-2	570/710	335	250	—	—	—	—	—	—	—	—
E360	Fe 690-2	670/830	360	250	—	—	—	—	—	—	—	—

1 For sections up to and including 100 mm only.
2 For wide flats and sections up to and including 63 mm and 100 mm respectively.
3 For wide flats up to and including 50 mm and for sections no limit is stated.
4 Only available up to and including 25 mm thick.
5 The steel grades S235 (Fe 360A), S275 (Fe 430A) and S355 (Fe 510A) appear only in the English language version (BS EN 10025) as non-conflicting additions and do not appear in other European versions.
6 Verification of the specified impact value is only carried out when agreed at time of enquiry and order.

Symbols used in BS EN 10025
S   = Structural Steel
E   = Engineering Steel
"235" "275" "355" = min YS (N/mm²) @ t ≤ 16 mm
JR = Longitudinal Charpy V-notch impacts 27J @ room temperature
JO = Longitudinal Charpy V-notch impacts 27J @ 0°C
J2 = Longitudinal Charpy V-notch impacts 27J @ -20°C
K2 = Longitudinal Charpy V-notch impacts 40J @ -20°C
G1 = Rimming steel (FU)
G2 = Rimming steel not permitted (FN)
G3 = Supply Condition 'N', i.e. normalized or normalized rolled
G4 = Supply Condition at the manufacturer's discretion.
Examples S235JRG1, S355K2G4

After BS EN 10025: 1993.

Table 3.4(b) *Structural Steels: Comparison between grades in BS EN 10113: 1993 and BS 4360: 1990*

BS EN 10113

Grade	UTS (N/mm²)	Min. YS at t = 16 mm (N/mm²)	Max. Thk (mm) (1)	(2)	Charpy (long) Temp. (°C)	Energy (J)	Max. Thk (mm) (1)	(2)
S275N	370 to 510	275	150	150	-20	40	150	150
S275NL					-50	27		
S355N	470 to 630	355	150	150	-20	40	150	150
S355NL					-50	27		
S420N	520 to 680	420	150	150	-20	40	150	150
S420NL					-50	27		
S460N	550 to 720	460	100	100	-20	40	100	100
S460NL					-50	27		
S275M	360 to 510	275	63	150	-20	40	63	150
S275ML					-50	27		

BS 4360: 1990

Grade	UTS (N/mm²)	Min. YS at t = 16 mm (N/mm²)	Max. Thk (mm) (1)	(2)	Charpy (long) Temp. (°C)	Energy (J)	Max. Thk (mm) (1)	(2)
43DD	430 to 580	275	-	100	-30	27	-	(7)
43EE			150 (3)	-	-50	27	75 (5)	-
50DD	490 to 640	355	150 (3)	100	-30	27	100 (5)	(7)
50E			-	100	-40	27	-	(7)
50EE			150 (3)	-	-50	27	75 (6)	-
-	-	-	-	-	-	-	-	-
-	-	-	-	-	-	-	-	-
55C	550 to 700	450	25	40	0	27	25	19
55EE			63 (4)	-	-50	27	63 (4)	-

Table 3.4(b)   *Continued*

BS EN 10113

Grade	UTS Min. YS at t = 16 mm (N/mm²)		Max. Thk (mm) (1)	(2)	Charpy (long) Temp. (°C)	Energy (J)	Max. Thk (mm) (1)	(2)
S355M	450 to 610	355	63	150	-20	40		
S355ML					-50	27	63	150
S420M	500 to 660	420	63	150	-20	40		
S420ML					-50	27	63	150
S460M	530 to 720	460	63	150	-20	40		
S460ML					-50	27	63	150

BS 4360: 1990

Grade	UTS Min. YS at t = 16 mm (N/mm²)	Max. Thk (mm) (1) (2)	Charpy (long) Temp. (°C)	Energy (J)	Max. Thk (mm) (1) (2)

Notes:
1 Applies to plates and wide flats.
2 Applies to Sections.
3 For wide flats max. thickness is 63 mm.
4 Not available as wide flats.
5 For wide flats max. thickness is 50 mm.
6 For wide flats max. thickness is 30 mm.
7 For sections no limit is given.

Symbols used in BS EN 10113
S = Structural Steel
'275' '355' '420' '460' = min YS (N/mm²) @ t ≤ 16 mm.
N = Normalized or normalized rolled
M = Thermomechanical rolled.
L = Low temperature (–50°C) impacts.
Examples S275N, S355ML

Longitudinal Charpy V-notch impacts

Grade	Min. ave. energy (J) at test temp (°C)						
	+20	0	-10	-20	-30	-40	-50
S_N/M	55	47	43	40			
S_NL/ML	63	55	51	47	40	31	27

After BS EN 10113: 1993.

Steel prices

The basis prices of BS 4360 steels, in the form of reversing mill plates and wide flats, are shown in Figure 3.11. It should be stressed that the prices shown in this figure were those in force on 1 March 1994 and it must be borne in mind that steel prices are adjusted from time to time. As indicated in Figure 3.11, the prices increase progressively at a given strength level as the impact properties are improved, i.e. from sub-grade A to sub-grade EE or F, and also as the strength level is increased, i.e. in moving from Grade 40 to Grade 55. However, as illustrated in Figure 3.12, the ratio of cost to yield strength falls very significantly as the strength of these steels is increased. Therefore there is a major cost incentive to utilize a higher strength grade in structures where the design is based primarily on yield strength.

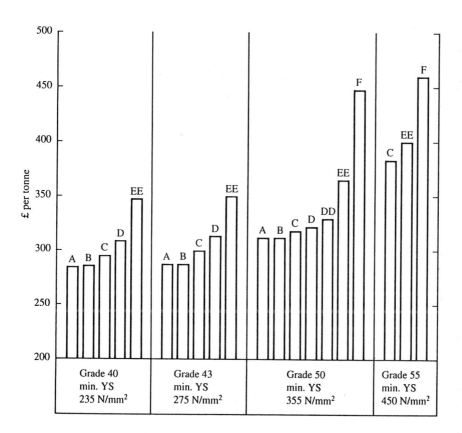

Figure 3.11 *Basis prices of reversing mill plates and wide flats as of 1 March 1994 (note: the price is dependent on thickness and the prices shown above are the lowest prices in each grade)*

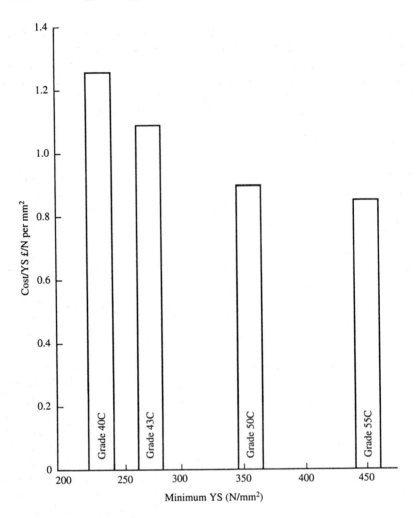

Figure 3.12 *Cost–yield strength relationships in BS 4360 Structural steels (plates and wide flats)*

Weathering steels

The term *weathering steels* is given to structural grades in which the resistance to atmospheric corrosion has been improved by the addition of small amounts of elements such as copper, phosphorus, silicon and chromium. These steels rust at a lower rate than plain carbon steels and, under favourable climatic conditions, they can develop a relatively stable layer of hydrated iron oxide which retards further attack. Very often, reference is made to the brown 'patina' that develops on the surface of these steels and the attractive appearance that it presents in buildings and bridges. However, regardless of their aesthetic qualities, weathering steels can provide cost savings by eliminating the initial painting operation and subsequent maintenance work.

These steels were first introduced in 1933 by the United States Steel Corporation under the brand name *Cor-Ten* and have been licensed for manufacture throughout the world. However, they have not been used extensively in the UK because frequent rain inhibits the formation of the stable oxide layer and rusting continues, albeit at a relatively slow rate. Ideally, long dry summer periods are required to develop the adherent oxide layer.

Corrosion resistance

The mid-specifications of the chemical composition in the Cor-Ten series are given in Table 3.5. Cor-Ten A is the original high-phosphorus grade whereas Cor-Ten B and Cor-Ten C were introduced in the 1960s. These later steels have normal levels of phosphorus and are micro-alloyed with vanadium to provide high strength.

The corrosion behaviour of these steels is shown in Figure 3.13[12] where they can be compared with plain carbon steel. Very clearly, the weathering grades are superior to carbon steel in each of the atmospheres investigated, Cor-Ten A providing a better performance than Cor-Ten B. However, in keeping with the corrosion behaviour of other types of steel, the performance is worst in marine atmospheres and weathering steels are not generally recommended for use in such environments.

In spite of the large amount of corrosion data that has been gathered on weathering steels, the mechanism of their superior performance is still relatively obscure. Studies by the United States Steel Corporation[13] showed that Cor-Ten steels rusted faster than carbon steel in the initial stages and it was only after a period of eight days that carbon steel showed a greater gain in weight. After that period, both materials had developed a continuous covering of rust mounds but those on the carbon steel grew to a larger size and eventually spalled from the surface. With Cor-Ten A, splitting was less frequent and no spalling was observed. X-ray diffraction work showed that the rust on both types of steel consisted essentially of γ-$Fe_2O_3.H_2O$ in the initial stages, but after about 30 days α-$Fe_2O_3.H_2O$ was detected. However, in addition to iron oxides, iron sulphates ($FeSO_4.3H_2O$, $FeSO_4.7H_2O$ and $Fe_2(SO_4)_3$) have been detected in the rust layers formed on steel in polluted atmospheres. It is suggested tentatively that the

Table 3.5

%	*Cor-Ten A*	*Cor-Ten B*	*Cor-Ten C*
C	0.08	0.14	0.16
Si	0.50	0.20	0.20
Mn	0.25	1.10	1.20
P	0.11	0.04 max.	0.04 max.
Cr	0.75	0.50	0.50
Ni	0.35	–	–
Cu	0.40	0.35	0.35
V	–	0.06	0.07

beneficial alloying elements render these sulphates less soluble and thereby retard the penetration of air and moisture through the oxide layer to the steel interface.

Horton[14] has examined the effect of individual alloying elements on the corrosion resistance of Mayari R steel (Bethlehem Steel Corporation). The base composition used in this work was as shown in Table 3.6.

Table 3.6

C%	Si%	Mn%	P%	S%	Cr%	Ni%	Cu%
0.08	0.28	0.70	0.10	0.03	0.60	0.40	0.60

The results are summarized in Figure 3.14(a), which shows the effect of variations in a single alloying element in the above base. The corrosion penetration of the base steel was 2.9 mm and is represented as a horizontal line. Copper was not examined in this investigation but Horton lists the following order of effectiveness for other elements:

Most beneficial	P
	Cr
	Si
	Ni
No effect	Mn
Detrimental	S

Horton also analysed data on Cor-Ten A steel which involved the base composition shown in Table 3.7. These results are shown in Figure 3.14(b) and

Table 3.7

C%	Si%	Mn%	P%	S%	Cr%	Ni%	Cu%
<0.10	0.22	0.25–0.40	0.10	<0.02	0.63	~0.5	0.42

(c) which deal with industrial and marine (Kure Beach, North Carolina) sites respectively. At the latter site, the order of effectiveness is as follows:

Most beneficial	P
	Si
	Cu (up to 0.3%)
	Cr
	Ni
	Cu (>3%)

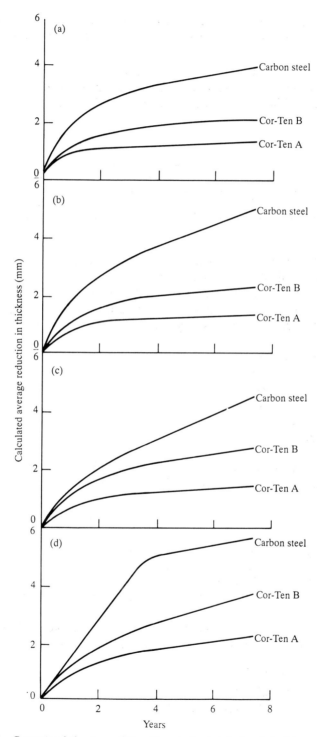

Figure 3.13 *Corrosion behaviour of Cor-Ten steels: (a) industrial; (b) semi-industrial; (c) semi-rural; (d) moderate marine*

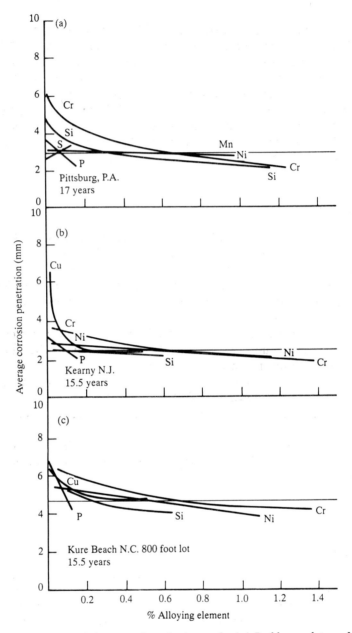

Figure 3.14   *Corrosion behaviour of weathering steels: (a) Bethlemen data on Mayari R; (b) USS Cor-Ten A in industrial atmosphere; (c) USS Cor-Ten A in marine atmosphere*

However, at the industrial site (Kearny, New Jersey), copper exerts a more powerful effect and would be promoted to a higher ranking than that shown above.

Very similar results have also been obtained by Hudson and Stanners[15] and Larrabee and Coburn[16] and it is generally acknowledged that phosphorus produces a marked and progressive benefit up to at least 0.1%. However, phosphorus is very detrimental to both toughness and weldability and therefore it is not included in some grades of weathering steel, with some loss in corrosion resistance. Copper is regarded as an essential constituent in weathering steels but little benefit is gained by increasing the copper content above about 0.3%. Elements such as silicon and chromium are mildly beneficial and the greatest benefit is obtained at levels up to about 0.25% and 0.6% respectively. As indicated earlier, manganese can be regarded as being neutral in its effect on corrosion resistance, whereas sulphur is detrimental.

Steel specifications

Weathering steels form part of the residue of grades in BS 4360: 1990 but are designated *weather-resistant steels*. Where appropriate, the mid specifications for the chemical composition of plate grades are given in Table 3.8. These steels are therefore very similar in composition to the Cor-Ten series, featuring a high-phosphorus grade (suffix A) and low-phosphorus-vanadium grades (suffixes B and C).

The tensile properties of the WR grades are similar to those of grade 50 steels with a minimum yield of strength of 355N/mm^2 in the smaller plate thicknesses. However, the impact properties are more limited, e.g. Grade WR50C offers a minimum of 27 J at $-15°C$ whereas the specified test temperature extends down to $-60°C$ in Grade 50F.

A European standard has now been introduced for these steels, namely BS EN 10155: 1993 *Structural steels with improved atmospheric corrosion resistance*. As shown in Table 3.9, this standard contains more grades than BS 4360: 1990, extending the range of both tensile and impact porperty requirements.

Table 3.8

%	WR50A	WR50B	WR50C
C	0.12 max.	0.19 max.	0.22 max.
Si	0.50	0.40	0.40
Mn	0.40	1.1	1.2
P	0.11	0.04 max.	0.04 max.
S	0.05 max.	0.05 max.	0.05 max.
Cr	0.85	0.6	0.6
Ni	0.65 max.	–	–
Cu	0.40	0.32	0.32
Al	–	0.03	0.03
V	–	0.06	0.06

Table 3.9   *Weathering Steels: comparison between grades in BS EN 10155: 1993 and BS 4360: 1990*

	BS EN 10155						BS 4360: 1990				
Grade[1]	Corresponding BS EN 10025 grade	UTS at t = 16 mm (N/mm²)	YS at t = 16 mm (N/mm²)	Charpy (long) Temp.	Charpy (long) Energy (J)		Grade	UTS at t = 16 mm (N/mm²)	YS at t = 16 mm (N/mm²)	Charpy (long) Temp.	Charpy (long) Energy (J)
S235J0W	S235J0	340 to 470	≥ 235	0	27						
S235J2W	S235J2G3			−20	27						
S355J0WP		490 to 630	≥ 355	0	27		WR50A(2)	≥ 480	≥ 345	0	27
S355J2WP				−20	27						
S355J0W	S355J0			0	27		WR50B(3)	≥ 480	≥ 345	0	27
S355J2G1W	S355J2G3			−20	27		WR50C(4)	≥ 480	≥ 345	−15	27
S355J2G2W	S355J2G4	490 to 630	≥ 355	−20	27		WR50C(4)	≥ 480	≥ 345	−15	27
S355K2G1W	S355K2G3			−20	40						
S355K2G2W	S355K2G4			−20	40						

Notes:

1 Plates and wide flats are available with speficied tensile and impact properties up to 100 mm thick, except grades S355J0WP and S355J2WP which are limited to 12 mm. Sections are available with specified tensile and impact properties up to 40 mm for all grades.

2 Avalable with specified tensile properties up to 40 mm thick and impact properties up to 12 mm thick.

3 Available with specified tensile and impact properties up to 50 mm thick.

4 Available with specified tensile properties up to 63 mm thick and impact properties up to 50 mm thick.

After BS EN 10155: 1993.

Symbols used in BS EN 10155

S   = Structural Steel

'275' '355' = min YS (N/mm²) @ t ≤ 16 mm

J0   = Longitudinal Charpy V-Notch impacts 27J @ 0°C

J2   = Longitudinal Charpy V-Notch impacts 27J @ −20°C

K2   = Longitudinal Charpy V-Notch impacts 40J @ −20°C

G1   = Supply condition 'N', i.e. normalized or normalized rolled.

G2   = Supply condition at the manufacturer's discretion.

W   = Weather resistant steel

P   = High phosphorus grade

Examples   S335J0WP, S355K2G2W

Clean steels and inclusion shape control

In parallel with developments in micro-alloy steels and thermomechanical processing, major attention has also been given to methods of producing steels which are isotropic with regard to tensile ductility and impact strength. Initially, the need for these steels was precipitated by the incidence of *lamellar tearing* in which plate material separates along planar arrays of non-metallic inclusions under the forces generated in highly restrained welds. However, the need for higher levels of impact strength in structural steels and the requirement for better cold formability in strip products have also focused attention on the development of cleaner steels and inclusion shape control.

The practice of adding calcium to steels for the reduction of sulphide and oxide inclusions is now used world-wide and has the added benefit of modifying the shape and size of these inclusions. One of the pioneers in this field was Thyssen Niederrhein in Germany and the practice is described by Pircher and Klapdar.[17] After tapping, calcium in the form of calcium silicide or calcium carbide is injected deep into the ladle by means of a refractory lined lance, using argon as a carrier. The calcium vaporizes and as it bubbles through the molten bath it combines with sulphur and oxygen in the steel. In either case, the reaction products are carried into the slag. The steel is generally deoxidized with aluminium prior to calcium treatment with an initial oxygen content in the range 20–100 ppm. After calcium treatment, the oxygen content is reduced to 10–20 ppm. The authors report that the desulphurizing effect of calcium is determined largely by the amount of calcium added but, as illustrated in Figure 3.15, the effect is influenced greatly by the type of ladle refractories

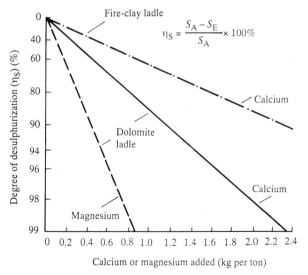

Figure 3.15 *Effect of calcium and magnesium on desulphurization. The desulphurizing additions were blown to a depth of 2.7 m in a steel bath having a temperature of 1580°C (After Pircher and Klapdar[17])*

employed. With dolomite ladles, little reaction takes place between the molten steel and refractories and the low level of oxygen in the steel enhances the effect of calcium as a desulphurizing agent. With dolomite ladles, the addition of 1 kg of calcium per ton reduces the sulphur content from an initial level of 0.02% to a final level of 0.003%. However, the authors state that sulphur contents less than 0.001% can be produced by this technique. As illustrated in Figure 3.15, magnesium is also very effective as a desulphurizing agent. However, calcium is generally preferred because it is cheaper and more controllable.

As indicated earlier, anisotropy of toughness and ductility is caused by the elongation of inclusions into planar arrays and both manganese sulphides and stringers of oxide are damaging in this respect. However, the problem is largely eliminated if the inclusions are present as small, isolated, non-deformed particles. Therefore major attention has been devoted to inclusion shape control in addition to the reduction in the volume fraction of inclusions. The elements used as inclusion modifiers are calcium, zirconium, tellurium and the rare earth metals.

In aluminium deoxidized steels, the inclusion population will generally include elongated Type II manganese sulphides, alumina and some silicates. However, after calcium treatment, the inclusions are restricted to calcium aluminate of the type $CaO.Al_2O_3$. The sulphur in the steel is also associated with these inclusions, either as calcium sulphide or as sulphur in solution. The calcium aluminate particles are globular in nature and tend to retain their shape on hot rolling. The beneficial effect of a reduction in sulphur content and calcium treatment on the reduction in area in the through-thickness direction is shown in Figure 3.16.

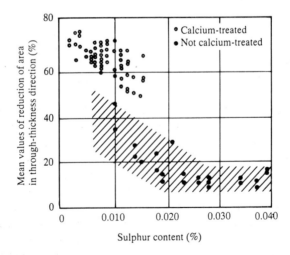

Figure 3.16 *Relationship between the sulphur content and the mean values of reduction of area of tensile test specimens in the through-thickness direction of steel grade FG 36 (After Pircher and Klapdar[17])*

Ships

Although there has been a significant decline in the world's output of new ships since the mid-1970s, the production of large merchant vessels remains a major application for structural steels. Despite marked changes in material selection for other forms of transport, steel remains virtually unchallenged for hull construction in large tankers and carriers. However, like other forms of steel construction, shipbuilding has witnessed the following changes since the early 1940s:

1. The change from riveting to welding as the principal method of joining.
2. An appreciation of the need for high levels of toughness.
3. The adoption of higher strength steels for reduced construction costs or higher operating efficiency.

Standard-strength steels

The main specifications for shipbuilding materials are issued by organizations known as *Classification Societies*, namely:

- American Bureau of Shipping
- Bureau Veritas
- Det Norske Veritas
- Germanischer Lloyd
- Lloyd's Register of Shipping
- Nippon Kaiji Kyokai
- Registro Italiano Navale

Whereas each of these societies publishes its own design rules and steel specifications, they collaborate closely through the International Association of Classification Societies (IACS). Therefore there is a high degree of uniformity in steel specifications in terms of composition, tensile properties and impact resistance. This collaboration stems from the mid-1950s when there was an urgent need to harmonize the approaches that had been taken individually in formulating steel specifications with improved resistance to brittle fracture.

Until the early 1940s, shipbuilding had been based on riveted construction and only one grade of steel was in common use, namely 'shipbuilding quality'. The steel was specified very simply in terms of tensile and bend tests and no limitations were placed on chemical composition. However, during World War II, the emergency shipbuilding programme in the United States demanded higher production rates for the so-called Liberty Ships and brought about the change from riveted to welded construction. This necessitated consideration of steel composition in relation to weldability but of far greater significance was the emergence of the problem of brittle fracture. In some instances, brittle fracture led to the catastrophic break-up of cargo vessels at sea and also produced the spectacular failure of the SS Schenectady in January 1943, whilst lying alongside the outfitting berth of a shipyard. Whereas it was shown that structural performance could be enhanced by means of improved design in critical elements, it was also evident that there was an urgent need to improve the toughness characteristics of ship plate steel.

In the years following World War II, each of the classification societies took independent action in formulating steel specifications, but by 1952 pressure from shipbuilders, owners and steelmakers brought about the initial discussions for the harmonization of specifications. A detailed account of these discussions was published by Boyd and Bushell[18] and at the outset, the seven classification societies collectively had a total of 22 grades of steel. However, these steels could be classified into three main types:

1. Ordinary ship steel, which was used in modest thicknesses and in lightly stressed areas.
2. An intermediate grade for areas where there was a need for some control over notch toughness and for intermediate thicknesses.
3. A high-grade steel with good notch ductility and for heavy plate thicknesses.

Although the seven societies recognized these three broad categories of steel, they were not able to rationalize their individual grades into three commonly acceptable specifications. On the one hand, the American Bureau of Shipping (ABS) favoured specifications based on deoxidation practice, composition and heat treatment whereas the European societies preferred specifications based primarily on mechanical properties. It was agreed finally to adopt unified grades based on both approaches and this resulted in five specifications for the three basic types of steel:

- Grade A – ordinary shipbuilding steel
- Grade B – intermediate grade based on the ABS approach
- Grade C – highest grade based on the ABS approach
- Grade D – intermediate grade based on specified impact strength at 0°C (European approach)
- Grade E – highest grade based on specified impact strength at −10°C (European approach)

Since that time, further rationalization has taken place and Lloyd's[19] now specifies four grades of steel with increasing impact strength requirements at the standard minimum yield strength value of 235 N/mm². Details of these steels are given in Tables 3.10. The yield strength requirement of these steels is identical to that specified for the lowest strength, Grade 40 steel in BS 4360 (Weldable structural steels) but, as illustrated later, the impact strength values are not completely compatible. Whereas normalizing was mandatory at one time for the higher toughness grades, these steels can now be supplied in the controlled-rolled condition, provided the specified mechanical properties are satisfied.

Table 3.10 *Lloyd's standard-strength shipbuilding steels*

(a) Chemical composition and deoxidation practice

Grade	A	B	D	E
Deoxidation	*Any method (for rimmed steel, see Note 1)*	*Any method except rimmed steel*	*Killed, see Note 2*	*Killed and fine grain treated with aluminium*
Chemical composition %				
Carbon	0.23 max.	0.21 max.	0.21 max.	0.18 max.
Manganese	*See* Note 3	0.8 min. ⎫ *See*	0.6 min.	0.7 min.
Silicon	0.5 max.	0.5 max. ⎰ Note 4	0.1–0.5	0.1–0.5
Sulphur	0.04 max.	0.04 max.	0.04 max.	0.04 max.
Phosphorus	0.04 max.	0.04 max.	0.04 max.	0.04 max.
Aluminium (acid soluble)	–	–	–	0.015 min. *See* Note 5

Carbon $+\frac{1}{6}$ of the manganese content is not to exceed 0.4%

Notes
1. For Grade A, rimmed steel may be accepted up to 12.5 mm thick inclusive, provided that it is stated on the test certificates or shipping statements to be rimmed steel and is not excluded by the purchaser's order.

2. Grade D steel may be supplied semi-killed up to 25 mm in thickness. In such cases, the requirement for the minimum silicon content does not apply.

3. For Grade A in thicknesses over 12.5 mm, the manganese content is to be not less than 2.5 times the carbon content.

4. For Grade B, when the silicon content is 0.1% or more (killed steel), the minimum manganese content may be reduced to 0.6%.

5. The total aluminium content may be determined instead of the acid-soluble content. In such cases the total aluminium content is to be not less than 0.02%.

(b) Mechanical properties for acceptance purposes

Grade	Yield stress (N/mm^2) minimum	Tensile strength (N/mm^2)	Elongation on $5.65 \sqrt{S_o}$ % minimum	Charpy V notch impact tests (longitudinal) Test temperature ($^\circ C$)	Charpy V notch impact tests (longitudinal) Average energy (J) minimum
A				–	–
B	235	400–490	22 (*see* Note 3)	0	27 (*see* Notes 2 and 4)
D				– 10	27 (*see* Note 2)
E				– 40	27 (*see* Note 2)

Notes
1. Requirements for products over 50 mm thick are subject to agreement.

2. For subsidiary impact test specimens the minimum average energy is to be:

Dimensions (mm)	Grades B, D and E
10 × 7.5	22 J
10 × 5	18 J

Where non-standard subsidiary test specimens are used, the minimum value is to be obtained by interpolation.

3. For full-thickness tensile test specimens with a width of 25 mm and a gauge length of 200 mm, the minimum elongation is to be:

Thickness (mm)		> 5 ⩽ 10	> 10 ⩽ 15	> 15 ⩽ 20	> 20 ⩽ 25	> 25 ⩽ 30	> 30 ⩽ 35	> 35 ⩽ 50
	⩽ 5							
Elongation (%)	14	16	17	18	19	20	21	22

4. Impact tests are generally not required for Grade B steel of 25 mm or less in thickness provided that satisfactory results are obtained from occasional check tests selected by the surveyor.

After Lloyd's Register of Shipping, *Rules and Regulations for the Classification of Ships.*[19]

Higher strength steels

By the mid-1960s, the higher strength steels, based on micro-alloy additions, had become established and each of the classification societies introduced specifications with yield stress values in the range 300–400 N/mm². The mechanical properties of high-strength steels currently specified by Lloyd's are shown in Table 3.11. The higher strengths are achieved by grain refinement and precipitation strengthening and the steels can be supplied in the as-rolled, controlled-rolled or normalized condition. The steels are normally made to a restricted carbon equivalent of 0.41% max., based on the formula:

$$C + \frac{Mn}{6} + \frac{Cr + Mo + V}{5} + \frac{Ni + Cu}{15} \ Wt\%$$

Whereas both BS 4360 and Lloyd's use the letters A to E to signify increasing levels of toughness, there are significant differences in the meaning of the designations, as shown in Table 3.12.

Most of the classification societies specify steels with minimum yield strength values of 315 and 355 N/mm² but Det Norske Veritas also lists a steel with a minimum yield strength of 390 N/mm².

Before concluding this section on steel specifications, brief mention should be made of the high-strength grades that are used in the construction of submarines. These can be considered as pressure vessels and, in order to withstand the very high hydrostatic pressures, submarine hulls are fabricated from quenched and tempered steels with minimum yield strength values of 550 N/mm² (Navy Q1) and 690 N/mm² (Navy Q2). Brief details of the compositions of these steels are given in Table 3.13.

Table 3.11 *Lloyd's higher strength shipbuilding steels. Mechanical properties for acceptance purposes*

Grade	Yield stress (N/mm^2) minimum	Tensile strength (N/mm^2)	Elongation on 5.65 $\sqrt{S_o}$ % minimum (see Note 2)	Charpy V notch impact tests (longitudinal)	
				Test temperature $(°C)$	Average energy (J) minimum (see Note 3)
AH32				0	31
DH32	315	440–590	22	− 20	31
EH32				− 40	31
AH34S				0	34
DH34S	340	450–610	22	− 20	34
EH34S				− 40	34
AH36				0	34
DH36	355	490–620	21	− 20	34
EH36				− 40	34

Notes
1. Requirements for products over 50 mm thick are subject to agreement.

2. For full-thickness tensile test specimens with a width of 25 mm and a gauge length of 200 mm, the minimum elongation is to be:

Thickness (mm)		$\leqslant 5$	> 5 $\leqslant 10$	> 10 $\leqslant 15$	> 15 $\leqslant 20$	> 20 $\leqslant 25$	> 25 $\leqslant 35$	> 35 $\leqslant 50$
Elongation (%)	Strength levels 32 and 34S	15	16	17	18	19	20	21
	Strength level 36	14	15	16	17	18	19	20

3. For subsidiary impact test specimens, the minimum average energy is to be:

Dimensions (mm)	Strength levels		
	32	34S	36
10 × 7.5	26	28	28
10 × 5	21	23	23

After Lloyd's Register of Shipping, *Rules and Requirements for the Classification of Ships.*[19]

Table 3.12

Grade designation	BS 4360	Lloyd's standard strength	Lloyd's higher strength
A	No test	No test	34 J at 0°C[a]
B	27 J at 20°C	27 J at 0°C	–
C	27 J at 0°C	–	–
D	27 J at −20°C	27 J at −10°C	34 J at −20°C[a]
E	27 J at −40°C	27 J at −40°C	34 J at −40°C[a]

[a]31 J min. for H32 grade steels.

Table 3.13

Grade	C%	Ni%	Cr%	Mo%	V%
Navy Q1	0.18 max.	2.75	1.4	0.4	
Navy Q2	0.13 max.	3.4	1.5	0.45	0.075

Because of their high alloy content, both compositions are capable of generating high strengths in thick-section plate after oil quenching. Excellent impact properties are also produced after tempering, the specifications calling for a minimum of 70 J at −84°C in plate thicknesses greater than 60 mm.

Design considerations

For design purposes, naval architects regard the hull of a ship as a beam or girder in which the deck and bottom form the flanges and the sides constitute the web. Lloyd's and the other classification societies specify the minimum thicknesses of plate or *scantlings* that shall be used in various parts of a ship which, when acting together, give the structure the required stiffness or *section modulus*. In general, the plate thicknesses are related to the length of a ship, assuming that the depth and breadth conform to a reasonably fixed ratio of its length. For example, the ratio of length to depth is not expected to exceed 16:1 and the length to breadth ratio is generally greater than 5:1.

Naval architects have to legislate for the worst situation that a ship is likely to encounter due to wave action, namely a wave with a length equal to that of the ship. As illustrated in Figure 3.17, this can give rise to two extreme conditions of stress:

1. Suspension at the middle position putting the deck in tension (*hogging*).
2. A wave at either end putting the bottom in tension (*sagging*).

Thus the area of the ship that is given greatest attention in design is the middle section or *midships*, since this is the area which is subjected to greatest stress and deflection. For this reason, the plate thicknesses in this region, designated *0.4L amidships*, are heavier than those required towards the ends of the ship. Between the two extremes, classification societies quote a taper in terms of percentage decrease in thickness per metre so as to avoid abrupt changes in section. The longitudinal strength of an I-beam is located in the flanges and the thickness of the deck and bottom plating are greater than those in the sides of a ship.

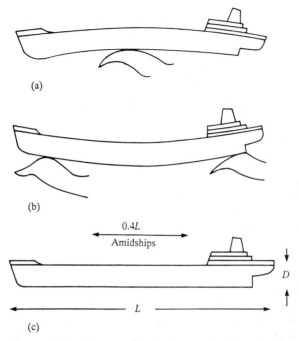

(a)

(b)

0.4*L*

Amidships

D

L

(c)

Figure 3.17 *Schematic illustration – hull deflection*

However, the longitudinal strength is not derived solely from the deck and bottom plating and stiffeners in the form of rolled sections (*bulb or plain flats*) are used to reinforce these areas and also the hull sides. Plating is also used in the longitudinal and transverse partitioning walls or *bulkheads*, which also contribute to the strength of a ship with respect to buckling.

For the most part, a ship's hull is constructed from Grade A steel but Lloyd's Rules distinguish between the material grade requirements for different parts of the hull. As shown in Table 3.14(a), five classes of material are identified in ascending order of fracture toughness which are translated into steel grades, according to thickness requirements, in Table 3.14(b). A schematic illustration of the location of Grades A, D and E in the midship section of a large tanker is shown in Figure 3.18. Thus the use of the tougher Grades D and E is confined mainly to the more highly stressed deck and bottom regions whereas Grade A is adequate for the sides of a vessel.

The use of higher strength steels in shipbuilding is attractive from two aspects:

1. Lower construction costs – from reduced steel weight and lower fabrication costs.
2. Lower operating costs – from reduced weight/lower fuel costs or higher carrying capacity for the same constructed weight.

Given the depressed state of the shipbuilding market, the former is probably the more important but the reduction in thickness that can be tolerated is governed by modulus and deflection considerations. From the analogy with a simple beam, the deflection of a ship is a function of the ratio of length to depth (L/D)

Table 3.14   *Steel selection in merchant ships*

(a)   Material classes

	Material class		
Structural member	*Within 0.4L amidships*	*Between 0.4L and 0.6L amidships*	*Outside 0.6L amidships*
Where $L > 250$ m: Sheerstrake or rounded gunwale Stringer plate at strength deck	V	III	II
Where $L \leqslant 250$ m: Sheerstrake or rounded gunwale Stringer plate at strength deck			
Bilge strake Deck strake in way of longitudinal bulkhead	IV	III	II
Strength deck plating Bottom plating including keel Continuous longitudinal members above strength deck Upper strake of longitudinal bulkhead Upper strake of topside tank	III	I	I
Deck plating, other than above, exposed to weather Side plating Lower strake of longitudinal bulkhead	II	I	I
External plating of rudder horn	–	–	III
Sternframe Internal components of rudder horn Rudder Shaft bracket	–	–	II

(b)   Steel grades

Thickness, t (mm)	Class									
	I		II		III		IV		V	
	Mild steel	*H.T. steel*	*Mild steel*	*H.T. steel*	*Mild steel*	*H.T. steel*	*Mild steel*	*H.T. steel*	*Mild steel*	*H.T. steel*
$t \leqslant 15$	A	AH	A	AH	A	AH	A	AH	D	DH
$15 < t \leqslant 20$	A	AH	A	AH	A	AH	B	AH	E	DH
$20 < t \leqslant 25$	A	AH	A	AH	B	AH	D	DH	E	EH
$25 < t \leqslant 30$	A	AH	A	AH	D	DH	E	DH	E	EH
$30 < t \leqslant 35$	A	AH	B	AH	D	DH	E	EH	E	EH
$35 < t \leqslant 40$	A	AH	B	AH	D	DH	E	EH	E	EH
$40 < t \leqslant 50$	B	AH	D	DH	E	EH	E	EH	E	EH
$t > 50$	B	AH	D	DH	E	EH	E	EH	E	EH

After Lloyd's Register of Shipping, *Rules and Regulations for the Classification of Ships.*[19]

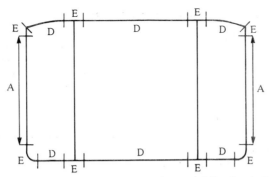

Figure 3.18 *Midships section showing distribution of Grades A, D and E*

and the design philosophy for the use of higher strength steels has been to limit deflection to that experienced in the most slender vessels in standard-strength steel. The reduction in thickness afforded by higher strength steels compared with that of standard-strength steel is given by the Lloyd's k factor:

$$k = \frac{235}{\sigma_0} \text{ or } 0.72 \text{ whichever is the greater}$$

where σ_0 is the minimum yield stress in N/mm^2.

In the case of standard-strength steel ($YS = 235 \ N/mm^2$), k is equal to 1.0 but decreases progressively with increasing yield strength, so permitting decreasing thicknesses of plate to be used compared with the standard grade. The lower limit of $0.72 \times$ thickness in standard steel ensures that deflection is maintained within reasonable bounds but in effect this means that no advantage will be gained, according to the present rules, by using a steel with a yield strength greater than 326 N/mm^2, i.e.:

$$\frac{235}{326} = 0.72$$

Therefore economic benefit is restricted currently to Lloyds H32 ($k = 0.75$) or H34 ($k = 0.69$) but the rules state that special consideration will be given to the use of steels with yield strengths greater than 355 N/mm^2.

The use of higher strength steels for shipbuilding is most advanced in Japan, where more than 50% of hull construction may be in high-strength steels. In particular, Japanese shipbuilders have taken advantage of accelerated cooled steels in which a yield stress of 345 N/mm^2 can be produced in a C–Mn steel with low carbon equivalent and excellent weldability. These steels are being used in icebreaking vessels since they provide high toughness in plate thicknesses up to 75 mm at temperatures of -60 to $-80°C$.

Offshore structures

Since the early 1970s, the UK has exploited its natural gas and oil resources and offshore platforms have become symbols of achievement in terms of design, materials and construction. One might therefore hold the view that offshore structures of this type were developed specifically for operation in the North Sea whereas such platforms were first constructed in the Gulf of Mexico in the 1940s. However, the conditions in the North Sea are considerably more severe with operating depths of 170 m compared to 7 m in the Gulf and with very much colder and rougher climatic conditions. Offshore platforms have now been constructed in large numbers and in 1981 it was reported[20] that more than 10 000 structures were in operation world-wide.

Although most offshore structures have been constructed in steel, several very large structures were built in reinforced concrete in the 1970s. This followed the tradition, established in the late 1880s, for the use of concrete in port and harbour installations, such as piers and jetties. Such facilities rank amongst the largest man-made structures and, for example, the Statfjord C platform had a float-out weight of more than 600 000 tonnes.[21] Concrete structures rely on their sheer mass to maintain their position on the seabed and for this reason are called gravity structures. Therefore they require good foundations and a thorough investigation of the seabed conditions to guard against long-term settlement and tilting. However, in spite of satisfactory performance, construction in reinforced concrete has remained relatively rare, apparently for economic reasons.

Design considerations

In the UK, the design and construction of offshore structures must comply with Guidance Notes[22] prepared by the Department of Energy. However, the Department of Energy has authorized the following organizations to issue *Certificates of Fitness* for offshore structures, in a similar manner to that for ships:

- American Bureau of Shipping
- Bureau Veritas
- Det Norske Veritas
- Germanischer Lloyd
- Offshore Certification Bureau
- Lloyd's Register of Shipping

The Guidance Notes illustrate very clearly the complex loading situation in offshore rigs by specifying the need to take account of the following loads in the primary structure:

- Dead loads – the weight of the structure and other fixed items
- Imposed or variable loads – loads from drilling operations, consumable stores, crew members, berthing and landing loads
- Hydrostatic loads – acting in a direction normal to the contact surface
- Environmental loads – wind, wave and current which act horizontally and, in extreme conditions, exert large overturning and sliding forces on fixed installations

Structural members in the splash zone are also subjected to major impact loads termed *slamming*. Whereas hydrostatic loads can be determined to a high degree of accuracy, environmental loads are random and the extreme magnitudes have to be predicted on a probability basis.

The most common type of steel platform is the welded tubular frame construction which is illustrated in Figure 3.19. In such platforms, the most critical areas are the *nodes* which represent the intersection of a *chord* (large-diameter member) and a *brace* (small-diameter member). Nodes can be constructed in various geometries, including T, K and T–K configurations, and they constitute areas of high stress concentrations. As such, they have a very marked effect on fatigue behaviour and therefore very careful consideration is given to the design and quality of welds in these locations. The first indication that fatigue could pose a major problem in offshore structures occurred in December 1965 when *Sea Gem*, a jack-up barge, collapsed in the North Sea, killing 13 people. It was shown that this failure was due to fatigue cracking

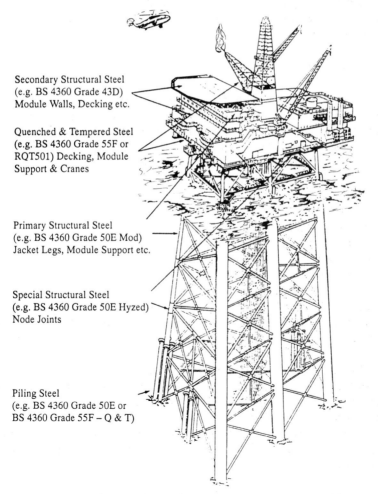

Secondary Structural Steel
(e.g. BS 4360 Grade 43D)
Module Walls, Decking etc.

Quenched & Tempered Steel
(e.g. BS 4360 Grade 55F or
RQT501) Decking, Module
Support & Cranes

Primary Structural Steel
(e.g. BS 4360 Grade 50E Mod)
Jacket Legs, Module Support etc.

Special Structural Steel
(e.g. BS 4360 Grade 50E Hyzed)
Node Joints

Piling Steel
(e.g. BS 4360 Grade 50E or
BS 4360 Grade 55F – Q & T)

Figure 3.19 *Types of steel for North Sea structures (After Billingham[21])*

which was probably initiated from brittle fracture, originating in welded components. Detailed consideration has therefore been given to the fatigue performance of nodal joints and design S–N curves have been produced which take account of the various stress concentration factors induced by different node geometries and plate thicknesses.

The Guidance Notes state that the calculated tensile stress in various members should not exceed 60% of yield stress under normal operating conditions and 80% of yield stress under extreme loading conditions.

Steel selection

Steels for tubular frame platforms can be grouped into three classes, depending on their location and duty:

1. Special structural steel.
2. Primary steel.
3. Secondary steel.

As illustrated in Figure 3.19, special structural steel is used for all the main nodes and also in the transition area to the shafts. This requires the highest grade of steel, namely BS 4360 Grade 50E Hyzed. As discussed earlier in this chapter, such steels are made to low sulphur contents and with alloy additions which modify the shape of the non-metallic inclusions so as to provide high levels of ductility in the through-thickness direction of the plate. Primary steel is used in all other structural members, such as jacket legs and topside module supports, and the steel is BS 4360 Grade 50E Mod. Secondary steel is used in lightly stressed areas such as module walls, decking, walkways and ladders and the specified material is BS 4360 Grade 43D.

Particular attention is paid to resistance to brittle fracture and the Guidance Notes lay down detailed impact and CTOD test requirements for parent plate, HAZ and weld materials. The background to these requirements has been published by Harrison and Pisarski[23] and involved a great deal of work on the correlation of Charpy V, wide plate and fracture toughness data. In the Guidance Notes, the impact test requirements are based on a minimum design temperature of $-10°C$. However, for other design temperatures, the Charpy test temperature is altered by 0.7°C for each 1°C that the design temperature differs from $-10°C$.

McLean and Oehrlein[24] have reported that the weight of steel involved in a typical tubular frame jacket is of the order of 13 000 tonnes and, together with the separately installed module support frame (2500 tonnes), supports the work area above sea level. The lower ends of the four external legs of the jacket penetrate through a *bottle* unit, each weighing 770 tonnes. These units transfer the service loads from the jacket to the piles which maintain the platform in position on the seabed. The topside modules, which house the equipment, work area and accommodation, weigh a total of 6600 tonnes. As indicated in a later chapter, austenitic stainless steels are now being used to clad the walls of the topside modules.

Cast steel nodes

In the early 1980s, significant effort was devoted to the development of cast steel nodes as alternatives to welded fabrications. The main technical incentive for this development was that castings could provide smooth, generous radii in these critical areas, compared with weld connections, and so reduce the stress concentration factor[25]. Typical compositions for these cast steel nodes are given in Table 3.15.

Table 3.15

	C%	Mn%	S%	Ni%	Nb%	V%	Mo%	Cr%	Cu%
Type I	0.14	1.30	0.005	0.43	0.025	0.05	0.11	0.09	0.04
Type II	0.15	0.85–1.70	0.01 max.	0.09–1.2					

According to Billingham,[21] cast steel nodes have enjoyed only limited commercial exploitation but in fact they were used at fatigue-prone locations in the Hutton Field.

Reinforcing bars

Large amounts of steel are used for the reinforcement of concrete in buildings, bridges and marine structures and reinforcing bars therefore constitute a competitor to structural steel plates and sections. At one time, reinforcing bars were regarded as low-grade, undemanding steel products and were often produced from diverted casts that were out of specification for the originally intended order. However, with the trend towards higher strength steels and the requirement for good fabrication characteristics, reinforcing steels are now made to high quality standards. Whereas some reinforcing bars are supplied in the form of plain carbon steel with a yield strength of 250 N/mm^2, extensive use is now made of higher strength steels with yield strengths up to 500 N/mm^2.

Standard specifications

The UK standard for steel reinforcement is BS 4449: 1988 *Carbon steel bars for the reinforcement of concrete*. It covers grades with minimum yield strength levels of 250 and 460 N/mm^2 in the form of plain rounds and *deformed* (ribbed) bars respectively. The chemical compositions of these grades are only broadly defined, as indicated in Table 3.16.

Table 3.16

	Grade 250	Grade 460
C%	0.25 max.	0.25 max.
S%	0.060 max.	0.050 max.
P%	0.060 max.	0.050 max.
N%	0.012 max.	0.012 max.

In order to provide a reasonable level of weldability, maximum carbon equivalent values of 0.42% (Grade 250) and 0.51% (Grade 460) must be observed, based on the following formula:

$$CE = C + \frac{Mn}{6} + \frac{Cr + Mo + V}{5} + \frac{Ni + Cu}{15} \; Wt\%$$

To ensure adequate ductility, minimum elongation values of 22% (Grade 250) and 12% (Grade 460) are required but the materials must also be capable of being bent through 180° around formers of the following proportions:

- Grade 250 – 2 × nominal bar diameter
- Grade 460 – 3 × nominal bar diameter

The specification also includes a rebend test requirement which was introduced initially to determine the susceptibility of the materials to strain age embrittlement. This involves bending bars initially through 45° around formers of the following proportions.

- Grade 250 – 2 × nominal bar diameter
- Grade 460 – 5 × nominal bar diameter

The bars are then immersed in boiling water for at least 30 minutes. On cooling to ambient temperature, the bars must be capable of being bent back towards their original shape through an angle of at least 23°.

Traditional reinforcing steels

In the UK, high-strength reinforcing steels have been produced traditionally by:

1. The cold twisting of plain carbon steels.
2. The use of vanadium-bearing micro-alloy steels.

At one time, the former was specified in a separate standard (BS 4461), but the cold-twisted product was incorporated in BS 4449 when this standard was revised in 1988. Although these steels employ different strengthening mechanisms, namely work hardening as opposed to precipitation strengthening, their properties are sufficiently similar to be covered by a single set of requirements for tensile and bend–rebend properties.

Whitely[26] reports that the cold-twisted product can be welded, with little loss of strength, by employing high heat inputs for short periods of time. Such practices restrict the area of the heat-affected zones and the quenching effect of adjacent material leads to the formation of strong, low-temperature transformation products. Whitely also indicates that the cold-twisted product provides adequate elevated-temperature tensile properties which are pertinent to the fire resistance of reinforced concrete structures.

Although not specified in BS 4449, the impact strength of reinforcing bars is of interest in relation to the potential damage that can be caused by the crashing of vehicles into the supporting columns and deck parapets of concrete road bridges. In collaboration with the Department of Transport, extensive work on this topic was carried out by British Steel Technical.[27] Using a large pendulum impact machine, with an impact velocity of 7.6 m/s, impact transition temperature curves were developed on full-section cold-twisted bars. It was shown that unnotched bars remained fully ductile at temperatures down to at least −65°C but, with the introduction of a sharp notch, the impact transition temperature was raised significantly. However, when impact tests were carried out on large reinforced concrete beams, the effect of notching the bars was shown to be significant only in the case of cold-twisted material. This is illustrated in Figure 3.20, which shows that very little loss of impact strength was observed in mild steel (Grade 250) bars and hot-rolled (micro-alloy, Grade 460) bars.

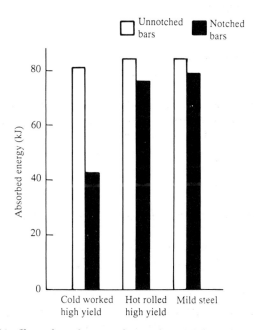

Figure 3.20 *The effect of notching reinforcing bars on the energy absorbed in concrete beams (After Armstrong et al[27])*

Controlled-cooled bars

In the mid-1970s, the CRM laboratories at Liege in Belgium published details of an in-line heat treatment process for the production of high-strength reinforcing bars.[28] Designated the *Tempcore* process, the application of controlled cooling after rolling results in the formation of an outer layer of martensite which is *temp*ered subsequently by conduction of heat from the *core* of the bars. The main aim of the process is to produce weldable, high-strength reinforcement cheaply by eliminating the costs associated with cold twisting or micro-alloy additions. The Tempcore process has been very successful, both technically and commercially, and has been licensed throughout the world.

The Tempcore process is illustrated schematically in Figure 3.21. On leaving the last finishing stand, the bar passes through a water cooling station which cools the outer region of the bar sufficiently quickly for the formation of martensite. At the end of the cooling process, the bar has an austenite core surrounded by a mixture of austenite and martensite, the amount of martensite increasing towards the surface of the bar. On leaving the cooling station, the bar is exposed to the atmosphere and the temperature gradient between the core and quenched surface begins to equalize. This leads to the tempering of the martensitic rim, providing an adequate balance between strength and ductility. During this second stage in the process, untransformed austenite in the outer layers of bars also transforms to bainite. The final stage of the process takes place as the bars lie on the cooling bed, namely the transformation of austenitic core. Depending upon factors such as composition, finishing temperature and

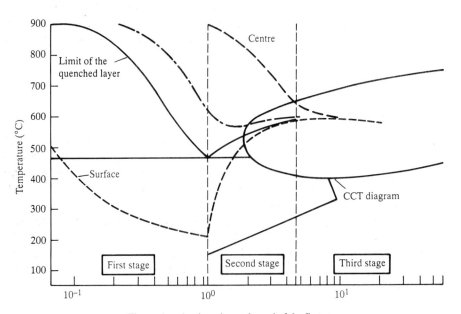

Time reduced to its value at the end of the first stage

Figure 3.21   *Tempcore process – thermal evolution in different parts of a 25 mm bar, finished rolled at 900°C (After Economopoulos et al.[28])*

cooling rate, the core can transform to ferrite and pearlite or to a mixed microstructure which includes some bainite. Thus the *Tempcore* process produces a variety of microstructures throughout the cross-section, ranging from tempered martensite in the outer layers to a region which is essentially ferrite and pearlite in the core, with an intermediate zone which may be predominantly bainitic.

Depending on the size of the bars, water cooling may be applied before the bars enter the finishing stands in order to reduce the length of the cooling station. This applies particularly to large-diameter bars which are finished at a high temperature and it is claimed[28] that the length of the cooling facility can be reduced by 70% by reducing the finishing temperature from 1050 to 900°C.

The Tempcore process is relatively cheap and provides the required mechanical properties in steels of low carbon equivalent. For these reasons, it is gradually superseding the cold-twisted and micro-alloyed products.

Steel bridges

In the UK, bridge construction is dominated currently by reinforced concrete and only about 15% of bridges are based on steel, compared to about 80% in Japan. According to Simpson,[29] one reason for this situation is that the design codes for steel bridges are complex and not easy to use. However, Simpson also points out that the failure during construction of three box girder bridges in the 1970s has also contributed to the lack of steel construction. In order to regain market share, steel suppliers and fabricators are attempting to improve the attraction of steel by providing cheaper and more effective forms of corrosion protection and by developing methods of construction that will allow faster erection schedules.

Design against brittle fracture

In the UK, the design and construction of bridges is covered by BS 5400: 1982 *Steel, concrete and composite bridges*. This code involves the use of the complex *limit state* approach for the calculation of design stresses and therefore this discussion on the use of steel in bridges will be confined to that part of the code dealing with avoidance of brittle fracture.

As a safeguard against brittle fracture, BS 5400: 1982 specifies the maximum thickness of steel that can be used in bridge tension members, with respect to the various grades of steel listed in BS 4360 (Weldable structural steels) and the minimum operating temperature of the bridge. The various stages in the determination of this temperature are outlined below:

1. The first stage is to determine the *minimum shade air temperature* for the location of the bridge from isotherm maps, based on Meteorological Office data.

2. This initial value is then adjusted for height above sea level by subtracting 0.5°C per 100 m. Additionally, there may be the need to take account of locations where the minimum temperatures diverge from published data, e.g. in frost pockets where the minimum may be substantially lower than the published value or in coastal and some urban areas where the minimum temperature may be higher.
3. The *minimum effective bridge temperature* (MEBT) is then derived from a table in which the minimum shade air temperature is adjusted to take account of the type of bridge construction, e.g. steel or concrete decking.
4. Finally, the *U* value for the bridge is determined by rounding down the MEBT to the next impact test temperature in BS 4360, i.e. a value of $-17°C$ would be rounded down to $-20°C$.

Each part of a bridge which is subjected to applied stress must be classified according to the following criteria:

- Type 1 – Any part subjected to an applied stress greater than 100 N/mm² and which has either
  (a) any weld connection
  (b) weld repair not subsequently inspected
  (c) punched holes not reamed.
- Type 2 – All parts subjected to applied stresses which are not of Type 1.

Stress calculations will have provided information on the combination of steel thickness and yield strength that will satisfy the required design strength and the appropriate sub-grade of steel in BS 4360 that provides the required level of impact strength can then be derived in two ways:

1. From a table in BS 5400 which provides a correlation between the limiting thickness for various grades and the *U* value – the minimum effective bridge temperature – differentiating between Type 1 and Type 2 stress conditions.
2. Directly from BS 4360, having calculated the required impact strength requirements in the following manner:

$$\text{for Type 1 } C_v \geqslant \frac{\sigma_y}{355} \frac{t}{2}$$

$$\text{for Type 2 } C_v \geqslant \frac{\sigma_y}{355} \frac{t}{4}$$

where C_v = Charpy V notch energy value in joules at the minimum effective temperature

σ_y = yield strength in N/mm²

t = thickness in mm

It should be noted that the divisor of 355 in the above equations corresponds to the minimum yield strength in N/mm² of BS 4360 Grade 50 steels.

Where severe stress concentrations occur, BS 5400 calls for more stringent toughness requirements and the impact energy value is calculated from:

$$C_v \geqslant \frac{\sigma_y}{355} \left[0.3t \, (1+0.67k) \right]$$

where k = stress concentration factor

Steel in multi-storey buildings

During the 1980s, there was a dramatic increase in the UK in the use of structural steelwork in multi-storey buildings, primarily at the expense of *in situ* concrete. The reasons cited[30] for this situation are:

1. The reduction in the price of structural steel relative to *in situ* concrete.
2. The reduced costs and improved methods of fire protection for structural steel.
3. The shorter site and total construction periods achieved through the use of steel.

All three aspects are cost or revenue related, the last item minimizing the period during which the capital for the investment has to be financed and decreasing the time at which the owner of a building begins to receive a return on investment.

In order to improve its share of the construction market, British Steel first analysed the contribution of the various items to the total cost of a steel building

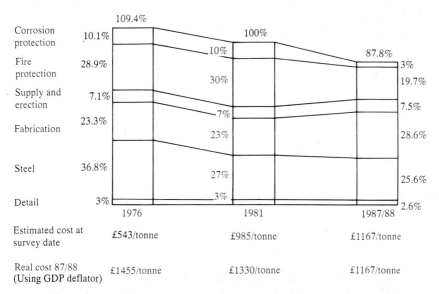

Figure 3.22 *Cost breakdown of multi-storey building frames (at 1987/88 prices)*

frame. This analysis is shown in Figure 3.22[31] and, whilst the cost of the basic steel was very significant, items such as corrosion protection, fire protection and fabrication also contributed very considerably to the overall cost. Each of these items was given detailed attention and, as illustrated in Figure 3.22, the cost of steel building frames has decreased significantly, in real value terms, over the 11–12-year period. Whereas the costs of corrosion protection and fire protection were reduced substantially, it should be noted that the reduction in the cost of steel played an even greater part in increasing the cost competitiveness of steel frames.

Building code requirements

In the UK, steel-framed buildings are specified in BS 5950: 1985 *Structural use of steelwork in buildings*. This refers to the use of structural steels specified in BS 4360 (Weldable structural steels) and the selection of a minimum level of toughness in relation to the yield strength of the steel, its thickness and service conditions.

The design strength P_y may be taken as $1.0 \times Y_s$ but cannot be greater than $0.84 \times U_s$ where Y_s and U_s are the minimum yield strengths and tensile strengths specified in BS 4360. The main types of steel used in building construction are Grades 43, 50 and 55 and, for convenience, BS 5950 tabulates the permissible design strengths for these grades, according to material thickness. This table is reproduced as Table 3.17.

Brittle fracture needs to be considered in locations which are subjected to tensile stresses in service. In this respect, the first step is the determination of the appropriate k factor, according to the level of tensile stress and location of material. This is determined from Table 3.18.

Table 3.17 *Design strength data for BS 4360 steels in building applications*

BS 4360 Grade	Thickness ($\leqslant mm$)	Design strength (N/mm^2)
43 A, B and C	16	275
	40	265
	100	245
50 B and C	16	355
	63	340
	100	325
55 C	16	450
	25	430
	40	415

After BS 5950: Part I: 1985.

Table 3.18

Tensile stress due to factored load at the location	Welded location	Unreamed punched holes	Non-welded location	Drilled or reamed holes
$\leqslant 100$ N/mm²	2	2	2	2
> 100 N/mm²	1	1	2	2

Having determined the thickness of material required from a consideration of design strength (Table 3.17), the required grade of steel can then be selected from a table in BS 5950, reproduced here as Table 3.19. This table includes the weathering grades (WR50 A, B and C) and differentiates between situations with k factors of 1 and 2, the limiting thickness/grade requirement being more severe when $k = 1$. This table also differentiates between the internal and external situations in buildings and applies providing the temperatures concerned do not fall below $-5°C$ and $-15°C$ respectively. When the steel is subjected to lower temperatures or where the steel grade or thickness is not covered in this table, then the toughness/grade requirements are determined by calculation. Thus the required impact strength at the service temperature is determined from:

$$C_v \geqslant \frac{Y_s t}{710\,k}$$

where $C_v =$ Charpy V notch energy in joules
$Y_s =$ the minimum yield strength of the material in N/mm²
$t \ =$ the thickness of the material in mm
$k \ =$ the factor determined from tensile stress/location considerations

Given that the design strength is based on yield strength, then substantial benefit is gained in building constructions from the substitution of carbon steel by higher strength structural steels, providing excessive deflection does not occur.

As illustrated in Figure 3.22, substantial savings have been made in the cost of corrosion protection and, in essence, this is related to the fact that paint systems were either over-specified or, in many cases, were simply not required.

In the previous section, it was stated that the minimum Charpy energy requirement for steel bridges was based on the following expression for Type 1 stress situations:

$$C_v \geqslant \frac{\sigma_y}{355}\left(\frac{t}{2}\right)$$

Apart from differences in nomenclature for yield strength and formula presentation, this expression is identical to that shown above for the minimum Charpy value for steel frame buildings, namely:

Table 3.19   *Selection of steels for buildings*

| BS 4360 Grade | Maximum thickness of parts subjected to applied tensile stress | | | |
| | Sections (except hollow sections) and flat bars | | Plates, wide flats and universal wide flats | |
	Internal (mm)	External (mm)	Internal (mm)	External (mm)
Values for k = 1				
43A	25	15	25	15
43B	30	20	30	20
43C	50	40	50	40
43D	50	50	75	75
43E	50	50	75	75
50A	16	10	16	10
50B	20	12	20	12
50C	40	27	75	55
50D	40	40	75	75
55C	19	16	19	16
55E	63	63	63	63
WR50A	12	12	12	12
WR50B	45	27	45	27
WR50C	50	50	50	50
Values for k = 2				
43A	50	30	50	30
43B	50	40	50	40
43C	50	50	50	50
43D	50	50	75	75
43E	50	50	75	75
50A	32	20	32	20
50B	40	25	40	25
50C	40	40	75	75
50D	40	40	75	75
55C	19	19	19	19
55E	63	63	63	63
WR50A	12	12	12	12
WR50B	50	50	50	50
WR50C	50	50	50	50

After BS 5950: Part 1: 1985.

$$C_v \geqslant \frac{Y_s t}{710\,k}$$

for situations where $k = 1$.

Steels for pipelines

Pipelines are very efficient for the mass transportation of oil and gas and can extend over vast distances. For example, the Alaskan National Gas Transportation System involves more than 4000 miles of linepipe in diameters ranging from 42 to 56 in. In Europe, the pipelines from northern Russia to Germany and Austria extend for more than 3100 miles and are predominantly in 56-in diameter linepipe. In either case, the operating pressures are up to 1450 lb/in². Over the past 30 years, the trend in pipeline design has been to larger sizes and higher operating pressures in order to increase the efficiency of transportation. This has been accomplished through the provision of steels with progressive increases in yield strength coupled with good weldability and sufficient toughness to restrict crack propagation. However, other trends in the extraction of oil and gas have imposed further requirements on the performance of linepipe steels. These include the need for high-strength, heavy-wall pipe to prevent buckling during pipeline installation in deep water, e.g. in depths of 170 m in the North Sea. The operation of pipelines in arctic regions, coupled with the transportation of liquid natural gas (LNG), have also imposed further demands for improved levels of toughness at low operating temperatures. However, a particularly important trend has been the exploitation of sour oil and gas reserves which has required the development of linepipe steels with resistance to hydrogen-induced cracking (HIC). Thus in addition to higher strength and toughness, developing pipeline technologies have required improved resistance to corrosion which has been met with specific alloy additions and special control over non-metallic inclusions. Of major importance in meeting the property requirements of the oil and gas industries have been the developments in the thermomechanical processing of steel and the bulk of linepipe steels are now supplied in the controlled-rolled condition.

Specifications and property requirements

Most linepipe specifications in the world are based on those issued by the American Petroleum Institute (API) which cover high test linepipe (5LX series) and spiral weld linepipe (5LS series). The API specification for high test linepipe was introduced in 1948 and at that time included only one grade, X42, with a yield strength of 42 ksi. Since that time, higher strength steels have been developed and the specification now includes grades up to X80 (80 ksi = 551 N/mm²).

Most specifications use yield strength as the design criterion, and as the yield strength is increased, the wall thickness can be reduced proportionately, using the same design safety factor. Thus the substitution of X70 by X80 grade can lead to a reduction in wall thickness of 12.5%, which illustrates very clearly the incentive for the use of higher strength steels. However, the measurement of yield strength in linepipe has been an area of contention between steelmakers and linepipe operators for some time because of the change in strength that occurs between as-delivered plate and pipe material. Tensile specimens are cut from finished pipe and these are cold flattened prior to testing. Due to the method of preparation, the yield strength measured in such test specimens can be significantly lower than that obtained on undeformed plate. The differential is due to the well established *Bauschinger effect* which leads to a decrease in yield strength when tensile testing is preceded by stressing in the opposite direction, e.g. during pipe unbending and flattening. In X70 pipe, the Bauschinger effect can result in a reduction in yield strength of 69 to 83 N/mm² (10 to 12 ksi) and therefore plate material has to be supplied with extra strength so as to compensate for this apparent loss in yield strength. However, when the yield strength of pipe material is measured in a ring tension test, the value is much closer to that measured in plate material. The Bauschinger effect in linepipe materials is particularly marked in traditional ferrite–pearlite steels which exhibit discontinuous yielding in the tensile test. The effect is reduced in steels containing small amounts of bainite or martensite, and in steels containing a significant amount of lower temperature transformation products, the unbending and flattening operation can lead to an increase in yield strength (or 0.2% proof stress) compared with undeformed plate. These materials exhibit continuous stress–strain curves and the high rate of work hardening compensates for the loss in strength due to the Bauschinger effect.

Toughness is a major requirement in linepipe materials and detailed consideration has been given to both fracture initiation and propagation. To design against fracture initiation, the concept of *flow stress dependent critical flaw size* is employed. This predicts the critical flaw size, relative to the Charpy toughness level, for specific pipe dimensions and operating pressures. Above this critical size of defect, the toughness level required to prevent fracture initiation becomes infinite and depends solely on the flow stress. The crack opening displacement test has also been used to determine fracture initiation in linepipe materials, particularly in relation to the heat-affected zone.

When fracture occurs in an oil pipeline, fluid decompression takes place very rapidly and therefore the driving force for crack propagation dissipates very quickly in time and space. However, this is not the case in gas pipelines and therefore the possibility of developing a long-running crack is a major concern. However, the avoidance of brittle propagation is generally assured by specifying a minimum of 85% shear area at the minimum service temperature, in full-thickness specimens in the Battelle Drop Weight Tear test. Various formulae have also been derived to specify the minimum Charpy level which will ensure the arrest of a propagating ductile fracture. These indicate that higher Charpy energy values are required for higher operating pressures and for pipes with higher strengths, larger diameters and heavier wall thicknesses. However, there still appears to be some concern about the adequacy of Charpy values for the prediction of crack arrest behaviour.

The weldability of linepipe materials is important, firstly in relation to pipe fabrication and secondly with regard to the girth welds that are used in the field for pipeline construction. Obviously the latter represent the more arduous welding requirements, particularly in low-temperature environments, but, in general, the use of carbon equivalent formulae appears to be adequate in ensuring crack-free welds. Traditionally, the International Institute of Welding (IIW) formula has been used to assess the weldability of materials:

$$\text{Carbon equivalent} = C + \frac{Mn}{6} + \frac{Cr + Mo + V}{5} + \frac{Cu + Ni}{15}$$

However, there is the general feeling that the IIW formula is not adequate to define the behaviour of modern steels with low carbon contents and the following relationship by Ito and Bessyo is sometimes preferred:

$$\text{Carbon equivalent} = C + \frac{Si}{30} + \frac{Mn + Cu + Cr}{20} + \frac{Ni}{60} + \frac{Mo}{15} + \frac{V}{10} + (B \times 5)$$

As indicated earlier, thermomechanical processing has permitted the development of high-strength steels with low carbon contents and this has contributed greatly to improved weldability in linepipe steels.

Linepipe manufacturing processes

Japan is a leading producer of linepipe and a summary of the processes and size ranges available in that country in 1981 is given in Figure 3.23. This indicates that linepipe is produced as seamless and welded tubing, the former being restricted to relatively small-diameter, thick-walled tubing. Welded pipes are produced by electric resistance welding (ERW) and submerged arc welding (SAW), the latter being used for both longitudinal and spiral welded pipe. ERW pipes are produced in sizes up to 600 mm (24 in) in diameter and up to 19 mm (0.75 in) wall thickness. Longitudinal welded pipes are produced mainly by the U-O process and account for most of the pipes used for oil and gas transmission lines. However, a small amount of longitudinal welded pipe is also produced by roll bending. As illustrated in Figure 3.23, very large diameter pipe is produced by the spiral welded process in wall thickness up to 25 mm (1 in).

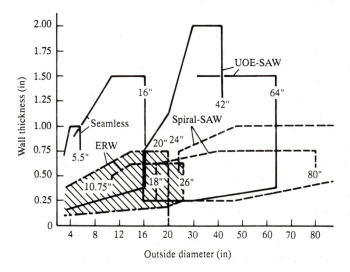

Figure 3.23 *Size ranges available with main Japanese linepipe manufacturing processes (After Nara et al.[32])*

Steel compositions for linepipe

A feature of the API 5LX specification is that it lays down very broad requirements for chemical composition, specifying only the maximum permitted levels for carbon, manganese, sulphur and phosphorus. On the other hand, customer specifications are much more restrictive in composition so as to obtain high levels of toughness and weldability at a specific level of yield strength. Even so, steelmakers can still exercise various options in terms of composition and thermomechanical processing and therefore a wide range of compositions is used to satisfy the property requirements of individual grades within the API specification. Sage[33] has produced a useful guide to the types of composition used for linepipe and this is shown in Figure 3.24. It will be noted that this chart differentiates between steels that exhibit the Bauschinger effect and those that experience a slight increase in yield strength during pipe making. The detailed chemical compositions used in various countries to satisfy API requirements are illustrated in Table 3.20. This includes some very low carbon, boron-treated steels[34] which develop a bainitic microstructure. In the production of these steels, the slab reheating temperature is in the range 1000–1150°C and the presence of fine particles of TiN inhibit austenite grain growth. The steels are finished at a temperature of about 700°C.

In general, X60 grade is satisfied by controlled-rolled ferrite–pearlite steels containing about 0.03% Nb. For X65 and X70, it is usual to supplement the grain-refining effect of niobium with dispersion strengthening from vanadium, and steels 3 and 5 in Table 3.20 are typical of this practice. In order to achieve X80 properties, small additions of nickel or molybdenum are made to the Nb–V steels and the steels are also subjected to a very severe controlled-rolling practice.

Pipeline fittings

Pipelines use a variety of fittings, such as valves, bends and tees, which have the same property requirements as linepipe, namely high strength, toughness and weldability. However, because of their method of production, the development of higher strengths in fittings has tended to lag behind that in linepipe material, which relies heavily on thermomechanical processing for the achievement of properties. Fittings are produced from forgings or as fabrications from plate and, in either case, the hot-working operations are not conducive to the generation or preservation of the fine-grained structure, characteristic of controlled-rolled linepipe. Pipeline fittings are therefore supplied mainly in the normalized or quenched and tempered condition. However, by limiting the hot-working temperature to about 750°C, fittings can be produced from controlled-rolled plate with little loss in properties.

Rogerson and Jones[39] have reviewed the topic of steels for high-pressure gas fittings and their summary of steel compositions is shown in Table 3.21. Thus, proprietary C–Mn–V steels, such as *Hyplus 29* and *Creuselso 42*, have been used for these applications which produce yield strength values up to 450 N/mm², depending upon section size. *Nicuage IN787* has been used successfully in the

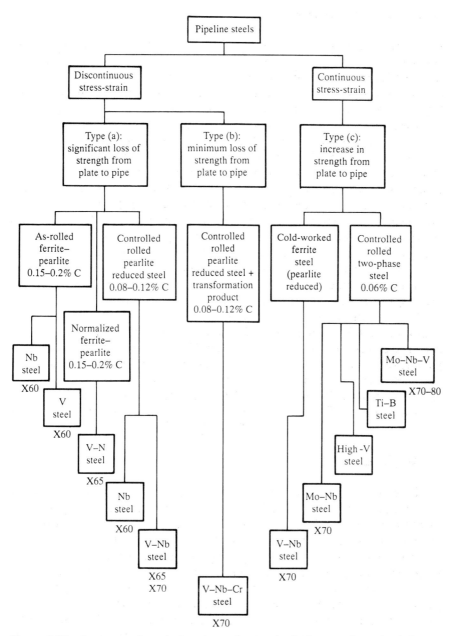

Figure 3.24 *Summary of steel pipes in production for high-strength steel pipelines in Europe, the Americas and Japan (After Sage[33])*

Alaskan oil pipelines in strength levels that satisfy X80. This Mn–Mo–Cu steel also has the type of composition that is resistant to sour gas/oil environments, as discussed in the next section.

Table 3.20 Chemical composition of linepipe steels

Steel no.	Grade	WT (mm)	C%	Si%	Mn%	P%	S%	Ni%	Mo%	Cu%	Nb%	V%	Ti%	B%	Ref.
1	X65	16	0.02	0.14	1.59	0.018	0.003	0.17			0.04		0.017	0.001	34
2	X65	25	0.03	0.16	1.61	0.016	0.003				0.05		0.016	0.001	34
3	X65	25	0.06		1.35	0.025	0.005	0.25		0.33	0.04	0.07			35
4	X70	20	0.03	0.14	1.91	0.018	0.003				0.05		0.018	0.001	34
5	X70	19.6	0.08		1.6						0.04	0.07			36
6	X80	12	0.07		1.65		0.002		0.22		0.05	0.075			37
7	X80	20	0.02	0.26	1.95	0.022	0.003	0.38	0.31		0.04		0.019	0.001	34
8	X80	19	0.08	0.1	1.5						0.052	0.076			38
9	X80	19	0.036	0.1	1.6			0.35	0.29		0.64				38

Table 3.21 Compositions of high-strength pipeline fittings

					Composition %											
		C	Si	Mn	S	P	Al	V	Nb	Ti	Ni	Cr	Mo	Cu	Co	N	Type
1	Hyplus 29	0.19	0.37	1.67	0.008	0.018	0.021	0.121	0.005	0.012	0.074	0.089	0.016	0.156	0.017	0.019	Tube
2	Controlled rolled	0.16	0.23	1.31	0.012	0.018	0.027	0.049	0.028	0.005	0.027	0.035	0.018	0.056	0.011	–	Plate
3	Controlled rolled	0.16	0.29	1.23	0.011	0.011	0.024	0.052	0.044	0.005	0.027	0.029	0.008	0.075		–	Bend
4	V–N–Ni steel	0.14	0.38	1.46	0.015	0.018		0.1			0.55	0.07	0.03	0.15		0.006	Plate
5(i)	V–N–Ni steel	0.19	0.4	1.63	0.004	0.008	0.01	0.108	0.005	0.005	0.55	0.166					Bend
(ii)	V–N–Ni steel	0.17	0.39	1.56	0.015		0.011	0.169		0.005	0.507	0.16	0.029	0.099		0.017	Bend
6	Acicular ferrite	0.06	0.26	1.59	0.009	0.022	0.043	0.018	0.029	0.005	0.213	0.005	0.341	0.008	0.005		Plate
7	Nicuage (IN787)	0.04	0.28	0.46	0.012	0.008	0.044	0.038	0.052	0.005	0.812	0.716	0.205	1.03	0.021	–	Plate
8	Modified Nicuage	0.05	0.26	1.23	0.008	0.015	0.027	0.031	0.071	0.005	1.1	0.018	0.206	1.18	0.012		Plate
9	BSC Mn–Mo–Cu	0.14	0.31	1.39	0.007	0.01	0.028	0.098	0.027	0.005		0.06	0.22	0.48		0.013	Plate

After Rodgerson and Jones.[39]

Steels for sour gas service

In recent years, the demand has increased for linepipe with resistance to environmental fracture due to the exploitation of sour gas wells, i.e. those containing significant levels of H_2S and CO_2. The National Association of Corrosion Engineers (NACE) has determined that a fluid is designated *sour* when it contains greater than 0.0035 atm partial pressure of H_2S. Both H_2S and CO_2 become corrosive in the presence of moisture and the corrosivity of natural gas is determined solely by the levels of these compounds. Although gas pipelines do not normally operate under corrosive conditions, a temperature drop in the gas to below its dewpoint or failure in dehydration plant can lead to the introduction of moisture. Similarly, other operational measures such as desulphurization and the use of inhibitors are not considered to be totally adequate and therefore there is a need for steels with inherent resistance to sour environments.

Two types of fracture can be introduced by H_2S, namely *hydrogen-induced cracking* (HIC) and *sulphide stress corrosion cracking* (SSCC). The latter form of attack is generally confined to steels with yield strengths greater than about 550 N/mm^2 and therefore does not feature prominently in linepipe steels. However, the fact that high hardness levels can be generated in the heat-affected zones of welds should not be overlooked since SSCC can result in catastrophic brittle fracture. On the other hand, HIC results in a form of blistering or delamination and can take place in the absence of stress. Atomic hydrogen is generated at cathodic sites under wet sour conditions and diffuses into the steel, forming molecular hydrogen at the interface between non-metallic inclusions and the matrix. When the internal pressure due to the build-up of molecular hydrogen exceeds a critical level, HIC is initiated. Long elongated inclusions such as Type II MnS are particularly favourable sites for crack initiation but planar arrays of globular oxides are also effective. Cracking can proceed along segregated bands containing lower temperature transformation products such as bainite and martensite. Cracking tends to be parallel to the surface but can be straight or stepwise.

The susceptibility to HIC is assessed by immersion of unstressed coupons in a synthetic solution of seawater, saturated in H_2S with a pH of 5.1–5.3 (BP test) or in the more aggressive solution of 0.5% CH_3COOH + 5% $NaCl$ + H_2S sat. at a pH of 3.5–3.8 (NACE test). In either case, the test duration is 96 h and the test parameters include crack length, crack width or blister formation.

From the remarks made earlier, it could be anticipated that the control of non-metallic inclusions would feature prominently in the development of HIC-resistant steels. The sulphur content is therefore generally reduced to below 0.01% and additions of calcium or rare earth metals are made to produce a globular sulphide morphology. In some Japanese practices, the sulphur contents are reduced to 0.001–0.003% with calcium additions of 0.0015–0.0035%.

Segregation effects can be minimized by restricting the levels of elements such as carbon, manganese and phosphorus and fortunately the use of controlled rolling enables high strengths to be obtained at low carbon levels. On the other hand, manganese is beneficial in improving toughness by refining the ferrite grain size and therefore a minimum level of manganese must be maintained, particularly in linepipe for arctic service.

Certain alloy additions are also effective in improving the resistance to HIC, notably copper and nickel. At pH levels above 5, these elements form protective films which prevent the diffusion of hydrogen into the steel. However, at pH levels below 5, these protective films are not formed and therefore these alloy additions show little advantage in the NACE test, outlined earlier. On the other hand, copper additions of 0.2–0.3% are extremely beneficial in preventing HIC, as determined by the BP test.

In summary, linepipe steels with resistance to HIC embody the following features:

1. Low levels of carbon, sulphur and phosphorus.
2. Additions of calcium or rare earth metals to globularize the sulphide inclusions.
3. Low levels of oxide inclusions.
4. Freedom from segregation so as to avoid bainitic or martensitic bands.
5. Additions of small amounts of copper or chromium.

References

1. Hall, E.O. *Proc. Phys. Soc. Series B*, **64**, 747 (1951).
2. Petch, N.J. *Proc. Swampscott Conf.*, MIT Press, p. 54 (1955).
3. Pickering, F.B. *Physical Metallurgy and the Design of Steels*, Applied Science Publishers.
4. Pickering, F.B. and Gladman, T. *ISI Special Report 81* (1961).
5. Irvine, K.J., Pickering, F.B. and Gladman, T. *JISI*, **205**, 161 (1967).
6. Gladman, T., Dulieu, D. and McIvor, I.D. In *Proc. MicroAlloying 75*, (Washington, 1975), Union Carbide, p. 25 (1975).
7. Vanderbeck, R.W. *Weld J.*, **37** (1958).
8. Sellars, C.M. In *Proc. HSLA Steels: Metallurgy and Applications* (eds Gray, J.M., Ko, T., Zhang Shouhua, Wu Baorong and Xie Xishan) (Beijing, 1985) ASM.
9. Tamura, I., Ouchi, C., Tanaka, T. and Sekine, H. *Thermomechanical Processing of High Strength Low Alloy Steels*, Butterworths (1988).
10. Cuddy, L.J. In *Proc. Thermomechanical Processing of Microalloyed Austenite* (eds Deardo, A.J., Ratz, G.A. and Wray, P.J.) (Warrendale Pa, 1982) Met. Soc. AIME (1982).
11. Cohen, M. and Hansen, S.S. In *Proc. HSLA Steels: Metallurgy and Applications* (eds Gray, J.M., Ko, T., Zhang Shouhua, Wu Baorong and Xie Xishan) (Beijing, 1985) ASM (1985).
12. United States Steel Technical Report – *Corrosion Performance of Mn–Cr–Cu–V Type USS Cor-Ten High Strength Low Alloy Steel in Various Atmospheres*, February 1965, Project No. 47.001-001 (7) and 47.001-003 (3).

13. United States Steel Technical Report – *A Study of the Initial Atmospheric Corrosion of Carbon Steel, USS Cor-Ten A Steel and USS Cor-Ten B Steel*, May 1966, Project No. 47.004-002 (1).
14. Horton, J.B. *The Rusting of Low Alloy Steels in the Atmosphere*, Paper to Pittsburgh Regional Technical Meeting of AISI, November 1965.
15. Hudson, J.C. and Stanners, J.F. *JISI*, July (1955).
16. Larrabee, C.P. and Coburn, S.K. In *Proc. First International Congress on Metallic Corrosion* (London, 1961) Butterworths (1961).
17. Pircher, H. and Klapdar, W. In *Proc. MicroAlloying '75* (Washington, 1975), Union Carbide Corporation (1977).
18. Boyd, G.M. and Bushell, T.W. *Trans. Royal Institution of Naval Architects*, July (1961).
19. Lloyd's Register of Shipping *Rules and Regulations for the Classification of Ships*, Part 3.
20. Graff, W.J. *Introduction to Offshore Structures*, Golf Publishing (1981).
21. Billingham, J. *Metals and Materials*, August, 472 (1985).
22. Department of Energy *Offshore Installations – Guidance on Design and Construction*, HMSO, London (1984).
23. Harrison, J.D. and Pisarski, H.G. *Background to New Guidance on Structural Steel and Steel Construction Standards in Offshore Structures*, HMSO, London (1986).
24. McLean, A. and Oehrlein, R. In *Proc. Performance of Offshore Structures*, Autumn Review Course Series 3, No. 7, The Institution of Metallurgists, p. 65 (1976).
25. Webster, S.E. *Journal of Petroleum Technology*, October, 1999 (1981).
26. Whitely, J.D. *Concrete*, December, 28 (1981) and January, 30 (1982).
27. Armstrong, B.M., James, D.B., Latham, D.J., Taylor, V.E., Wilson, C. and Heighington, K. In *Proc. First International Conference on Concrete for Hazard Protection* (Edinburgh, 1987) The Concrete Society, p. 241 (1987).
28. Economopoulos, M., Respen, Y., Lessel, G. and Steffes, G. *Application of Tempcore Process to the Fabrication of High Yield Strength Concrete – Reinforcing Bars*, CRM No. 45, December 1975.
29. Simpson, R.J. *Metals and Materials*, October, 598 (1989).
30. Constructional Steelwork Economic Development Committee *Efficiency in the Construction of Steel Framed Multi-Storey Buildings*, National Economic Development Office.
31. Preston, R.R. Private communication.
32. Nara, Y., Kyogoku, T., Yamura, T. and Takeuchi, I. In *Proc. Steels for Line Pipe and Pipeline Fittings* (London, 1981) The Metals Society/The Welding Institute, p. 201 (1981).
33. Sage, A.M. In *Proc. Steels for Line Pipe and Pipeline Fittings* (London, 1981) The Metals Society/The Welding Institute, p. 39 (1981).
34. Nakasugi, H. *et al.* In *Proc. Steels for Linepipe and Pipeline Fittings* (London, 1981) The Metals Society/The Welding Institute, p. 94 (1981).
35. Cavaghan, N.J. *et al.* In *Proc. Steels for Linepipe and Pipeline Fittings* (London, 1981) The Metals Society/The Welding Institute, p. 200 (1981).

36. Shiga, C. *et al.* In *Proc. Steels for Linepipe and Pipeline Fittings* (London, 1981) The Metals Society/The Welding Institute, p. 134 (1981).
37. Coolen, A. *et al.* In *Proc. Steels for Linepipe and Pipeline Fittings* (London, 1981) The Metals Society/The Welding Institute, p. 209 (1981).
38. Lander, H.N. *et al.* In *Proc. Steels for Linepipe and Pipeline Fittings* (London, 1981) The Metals Society/The Welding Institute, p. 146 (1981).
39. Rogerson, P. and Jones, C.L. In *Proc. Steels for Line Pipe and Pipeline Fittings* (London, 1981) The Metals Society/The Welding Institute, p. 271 (1981).

4 *Engineering steels*

Overview

The term *engineering steels* applies to a wide range of compositions that are generally heat treated to produce high strength levels, i.e. tensile strengths greater than 750 N/mm². These steels are subjected to high service stresses and are typified by the compositions that are used in automotive engine and transmission components, steam turbines, bearings, rails and wire ropes. As well as carbon and low-alloy grades, engineering steels also embrace the *maraging* compositions that are generally based on 18% Ni and which are capable of developing tensile strengths greater than 2000 N/mm².

Engineering steels are concerned primarily with the generation of a particular level of strength in a specific section size or *ruling section*. This introduces the concept of *hardenability* which is concerned with the ease with which a steel can harden in depth rather than the attainment of a specific level of hardness/strength. In turn, this relates to the effects of alloying elements on hardenability and the influence of cooling rate on a specific composition or section size. Much of the information that is available today on hardenability concepts and the metallurgical factors affecting hardenability was generated in the United States in the 1930s with names such as Grossman, Bain, Grange, Jominy and Lamont featuring prominently in the literature. This period also coincided with the introduction of isothermal transformation diagrams which paved the way to the detailed understanding of the decomposition of austenite and a qualitative indication of hardenability.

Up until the late 1940s, engineering steels often contained substantial levels of nickel and molybdenum, the concept being that these elements were required in order to provide a good combination of strength and toughness. Whereas these alloy additions certainly fulfilled this objective, what was to change in subsequent years was the generation of quantitative data on the actual level of toughness that was required in engineering components. This paved the way to the substitution of nickel and molybdenum by cheaper elements such as manganese, chromium and boron and the more economical use of alloy additions for particular hardenability requirements. The theme of cost reduction was also pursued very vigorously in the 1970s and 1980s with the introduction of medium-carbon, micro-alloy steels for automotive forgings. As illustrated later in the text, these steels offer the potential of major savings over traditional quenched and tempered alloy grades through lower steel costs, the elimination of heat treatment and improved machinability.

Steel cleanness and the reduction of non-metallic inclusions have also been of major concern to users of engineering steels, particularly in applications with the potential for failure by fatigue. Bearing steels are a typical example and the fatigue performance of these steels has been improved progressively over the years with the adoption of facilities such as vacuum degassing (1950s), argon shrouding of the molten stream (1960s) and vacuum steelmaking (1970s). However, major improvements in cleanness have also been obtained in bulk steelmaking processes for bearing steels with the introduction of secondary steelmaking facilities.

The machining of automotive components can account for up to 60% of the total cost and therefore major effort has been devoted to the development of engineering steels with improved machinability. In the main, these developments have been focused on the traditional resulphurized grades but with the addition of elements such as calcium and tellurium for sulphide shape control and improved transverse properties.

In summary, the author's overall perception of this sector has been one of continuing effort to achieve cost reduction, initially through the use of cheaper alloying elements but latterly via the concept of lower-through costs and involving a reduction in the cost of heat treatment and machining.

Heat treatment aspects

Isothermal transformation diagrams

Isothermal transformation diagrams were first published by Bain and Davenport[1] in the United States in 1930 and paved the way to the detailed understanding of the effects of alloying elements on the heat treatment response in steels. A steel is first heated to a temperature in the austenitic range, typically 20°C above Ac_3, and then cooled rapidly in a bath to a lower temperature, allowing isothermal transformation to proceed. The progress of transformation can be followed by dilatometry, the degree of transformation depending upon the holding time at temperature. As illustrated schematically in Figure 4.1, the

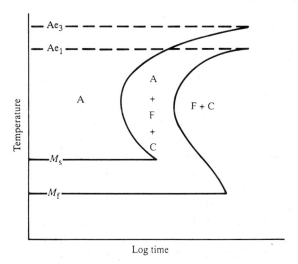

Figure 4.1 *Isothermal transformation diagram*

start and finish of transformation to ferrite, pearlite, bainite and martensite are then shown on a diagram as a function of temperature and time. Isothermal transformation diagrams, also known as TTT (time-temperature-transformation) diagrams, are simple in concept but are not representative of the majority of commercial heat treatments which involve a continuous-cooling operation. However, *martempering* and *austempering* are examples of heat treatment that employ isothermal sequences.

A schematic illustration of martempering is given in Figure 4.2. Following a conventional austenitizing treatment, a component is cooled rapidly to a temperature just above the M_s temperature and held at this temperature long enough to ensure that it attains a uniform temperature from surface to centre. The material is then air cooled through the M_s–M_f temperature range to form martensite and is subsequently tempered. Martempering is therefore not a tempering operation but a treatment that leads to low levels of residual stresses and minimizes distortion and cracking.

In *austempering*, Figure 4.3, a component is quenched into a salt bath, again at a temperature above M_s, but the material is held at this temperature long enough to allow complete isothermal transformation to bainite. The component is then quenched or air cooled to room temperature. Again no tempering is involved in the conventional sense but austempering represents a short, economical heat treatment cycle that provides a good combination of strength and toughness. The cooling rate must be fast enough to avoid the formation of pearlite and the isothermal treatment must be long enough to complete the transformation to bainite.

Although they proved to be extremely useful in gaining a better understanding of austenite decomposition and transformation kinetics, it became apparent that isothermal transformation diagrams were only of limited value in the evaluation of continuous-cooling processes. Although attempts were made to convert isothermal transformation diagrams to a continuous-cooling format,

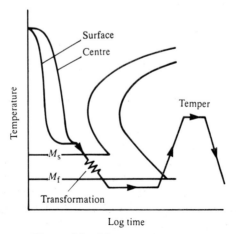

Figure 4.2   *Martempering treatment*

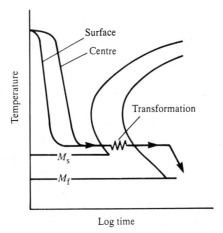

Figure 4.3   *Austempering treatment*

these were only partially successful. Attention turned therefore to the direct determination of continuous-cooling transformation diagrams.

Continuous-cooling transformation (CCT) diagrams

CCT diagrams are generated from a series of temperature–length curves. A specimen is heated slowly into the austenitic range and, as illustrated in Figure 4.4, the heating curve provides the facility for the determination of the Ac_1 and Ac_3 temperatures. The specimen is then cooled at a prescribed rate and the start and finish of transformation can be determined respectively from the initial deviation from the cooling curve and the subsequent conformity to the heating curve. This exercise is repeated for series of cooling rates, ranging from the simulation of water quenching in a small-diameter rod to that experienced in furnace cooling.

CCT diagrams can be presented in two ways:

1. Temperature–time plots in which the cooling time is plotted horizontally on a log scale.
2. Temperature–bar diameter plots, the latter representing different bar sizes cooled at rates simulative of air, oil and water cooling.

Examples of both types of presentation are shown in Figures 4.5 and 4.6. A feature of the temperature–time presentation is the insertion of the cooling curves used to generate the CCT diagrams and the hardness developed after each cooling rate is shown in a circle at the end of the cooling curves. This type of presentation was favoured by Cias[2] of Climax Molybdenum who compiled a series of CCT diagrams for medium-carbon alloy steels. Temperature–bar diameter plots were used by Atkins[3] of British Steel in a major publication of CCT diagrams, covering a wide range of carbon and alloy steels.

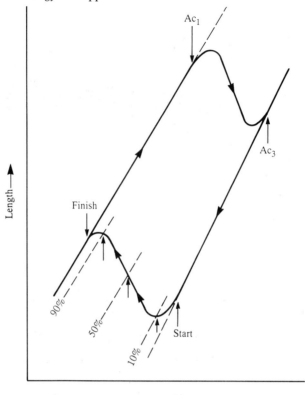

Figure 4.4 *Length changes in heating and cooling*

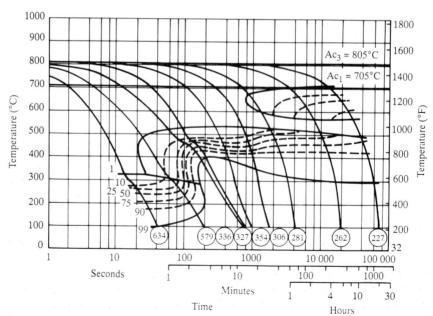

Figure 4.5 *Continuous-cooling transformation diagram for a 0.39% C 1.45% Mn 0.49% Mo steel (After Cias[2])*

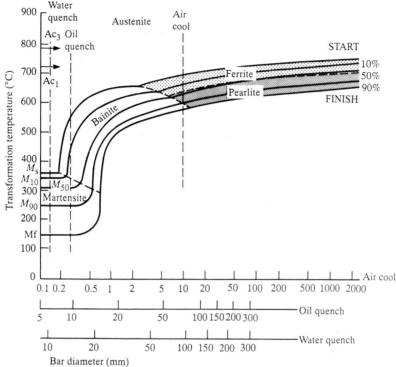

Figure 4.6   *Continuous-cooling transformation diagram for a 0.38%C 0.6% Mn steel; the dotted vertical lines indicate the transformations that occur in a 10 mm bar after air cooling, oil quenching and water quenching respectively (After Atkins[3])*

Hardenability testing

The term *hardenability* is used in various ways to describe the heat treatment response of steels, employing one or other of interrelated parameters such as hardness and microstructure. When evaluated by hardness testing, hardenability is often defined as the capacity of a steel to harden *in depth* under a given set of heat treatment conditions. What must be emphasized in this definition is the fact that hardenability is concerned with the depth of hardening or the hardness profile in a component, rather than the attainment of a specific level of hardness. Using microstructure as the control parameter, Siebert *et al.*[4] have defined hardenability as:

'the capacity of a steel to transform partially or completely from austenite to some percentage of martensite at a given depth when cooled under some given condition'

Whereas the hardenability of a steel can be determined from continuous-cooling transformation diagrams, the construction of such diagrams is both time-consuming and expensive and therefore more economical methods were required for the measurement of hardenability. A number of hardenability tests were developed in the 1930s but the best known and most widely used is the

Jominy end quench test. Developed in imperial units, the Jominy specimen is a cylinder 102 mm (4 in) long × 25.4 mm (1 in) diameter with a flange at one end. It is usual to normalize the material to be tested prior to machining the specimen in order to eliminate variations in the hardening response that might be introduced by differences in microstructure in the as-rolled condition. The specimen is heated to the appropriate austenitizing temperature and then transferred quickly to a fixture which suspends the specimen above a tube through which a column of water is directed against the bottom face. The arrangement is shown in Figure 4.7 and it should be emphasized that the water flow is tightly specified and controlled in order to produce a consistent quenching effect. Whereas the quenched end of the specimen experiences a rapid rate of cooling, the effect diminishes along the length of the specimen to give a value approaching air cooling at the other end. When the quenching operation is complete, flats are ground at diametrically opposed positions on the specimen to a depth of 0.38 mm to remove decarburized material and provide a suitable surface for hardness testing. This can involve either Vickers (HV) or Rockwell (HRC) hardness testing at intervals of about 1.5 mm for alloy steels or 0.75 mm for carbon steels. A typical Jominy hardenability curve is shown in Figure 4.8, which reflects the martensitic hardening that can be achieved at the end quenched position in a steel of reasonable hardenability and transformation to bainite and ferrite–pearlite at the slower rates of cooling.

Although it remains predominant, the Jominy hardenability test has been criticized for not being sufficiently discriminating between steels of low hardenability, i.e. those involving a very rapid decrease in hardness just beyond the quenched end. For such steels, the *SAC* test is deemed to be more appropriate, although it has been used very rarely in the UK. In this test, the specimen is again a cylinder but measures 140 mm long × 25.4 mm diameter. After normalizing and austenitizing at a suitable temperature above Ac_3, the specimen

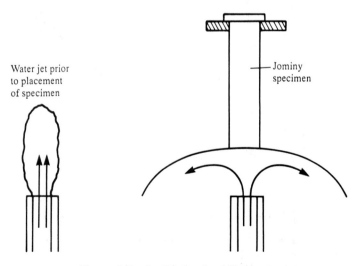

Water jet prior to placement of specimen

Jominy specimen

Figure 4.7 *Jominy hardenability testing*

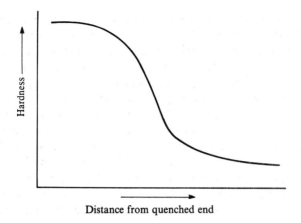

Figure 4.8 *Schematic Jominy hardenability curve*

is quenched overall in water. After quenching, a cylinder 25 mm long is cut from the test specimen and the end faces are ground very carefully to remove any tempering effects that might have been introduced during cutting. Rockwell (HRC) hardness measurements are then made at four positions on the original cylinder face and the average hardness provides the surface (S) value. Rockwell testing is then carried out along the cross-section of the specimen from surface to centre and provides the type of hardness profile illustrated in Figure 4.9. The total area under the curve provides the area (A) value in units of Rockwell-inch (using the original imperial unit) and the hardness at the centre gives the C value. The SAC value of a steel might be reported as 65-51-39, which would indicate a surface hardness of 65 HRC, an area value of 51 Rockwell-inch and a

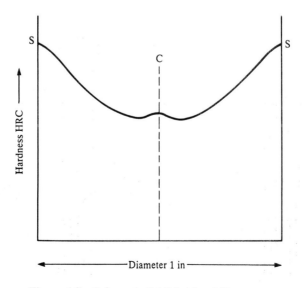

Figure 4.9 *Schematic SAC hardenability curve*

centre hardness of 39 HRC. One interesting feature of the SAC test is that it can reveal central segregation in a bar as indicated by a hardness peak at the centre position.

Factors affecting hardenability

Grain size

In a homogeneous austenitic structure, the nucleation of pearlite occurs almost exclusively at the grain boundaries and therefore the larger the grain boundary surface area, the greater are the nucleation sites for pearlite formation. Thus the hardenability of a given composition will increase with increasing austenitizing temperature and austenite grain size. The major effect of austenitizing temperature on the hardenability of a 0.55% C 1% Cr 0.2% Mo steel is shown in Figure 4.10, which is based on the work of Grange and presented by Grossmann.[5]

Although grain coarsening could be employed as a cheap method of achieving high hardenability, this approach is rarely adopted because the toughness and ductility are impaired. Instead, most commercial engineering steels are made to an aluminium-treated, fine-grain practice in order to produce microstructures that will provide a good combination of strength and toughness/ductility.

Alloying elements

Because of their very distinct effects on hardenability, it is convenient to consider alloying elements in three separate groups:

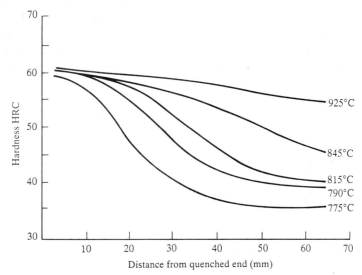

Figure 4.10   *Effect of austenitizing temperature on the Jominy hardenability of 0.55% C 0.84% Mn 0.92% Cr 0.21% Mo steel (After Grange[5])*

1. Carbon.
2. General group – Cr, Mn, Mo, Si, Ni, V etc.
3. Boron.

Carbon must be placed in a special category because it is the element that controls the hardness of martensite and therefore defines the maximum hardness that can be achieved in a given steel composition. The relationship between carbon content and the hardness of martensite is shown in Figure 4.11. The effect is reasonably linear at carbon contents up to about 0.6% but, depending upon the alloy content of the steel, the hardness then reaches a plateau and declines at higher carbon contents. This is due to the depression of the M_f temperature below room temperature, leading to incomplete transformation and the presence of retained austenite. However, because of the marked effect of carbon on hardness, it is important to bear this point in mind when attempting to compare the hardenability of steels of different carbon content.

In addition to its effect on hardness, carbon also has a significant effect on hardenability. This is illustrated in Figures 4.12 and 4.13, which show the progressive effect of carbon in base steels containing 0.8% Mn and 0.5% Ni 0.5% Cr 0.2% Mo (SAE 8600) respectively.[6] The effect of carbon on the hardness of martensite is evident in both types of steel but the effect on hardenability is very much more pronounced in the Ni–Cr–Mo steel. However, carbon is rarely used as a hardenability agent because of its adverse effect on toughness and its tendency to promote distortion and cracking. In addition, high-carbon steels are hard and difficult to cut or shear in the annealed condition.

With the exception of cobalt, small additions of all alloying elements will retard the transformation of austenite to pearlite and thereby increase hardenability. However, the elements that are most commonly used for the promotion of hardenability are manganese, chromium and molybdenum but nickel and vanadium are frequently incorporated for additional purposes. A considerable amount of work has been carried out, particularly in the United States, to quantify the effects of the major alloying elements on hardenability and, despite some complex and interactive effects, the general order of potency has been

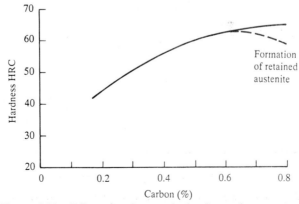

Figure 4.11   *Effect of carbon on the hardness of martensite*

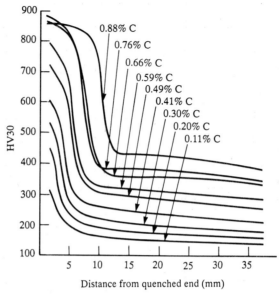

Figure 4.12   *Effect of carbon on hardenability of a 0.8% Mn steel (After Llewellyn and Cook⁶)*

established. The published data on this topic have been reviewed very thoroughly by Siebert *et al.*,[4] who include information on the effects of elements such as copper, tungsten and phosphorus as well as the five common alloying elements.

Work by deRetana and Doane[7] evaluated the effects of the major alloying elements on the hardenability of low-carbon steels of the type used for case carburizing. Their information is shown in Figure 4.14, where the change in hardenability is expressed by means of a multiplying factor, calculated as follows:

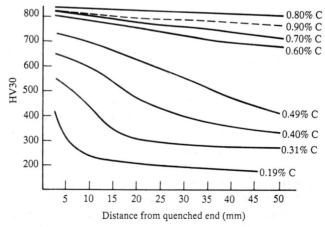

Figure 4.13   *Effect of carbon on hardenability of SAE 8600 steels (After Llewellyn and Cook⁶)*

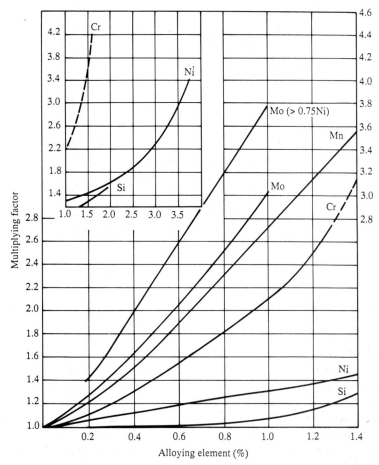

Figure 4.14 *Average multiplying factors for several elements in alloy steels containing 0.15–0.25% carbon (After deRetana and Doane[7])*

$$\text{Multiplying factor} = \frac{\text{hardenability of (base steel + alloying element)}}{\text{hardenability of base steel}}$$

The data shown in Figure 4.14 were derived from Jominy tests on a variety of commercial and experimental casts of carburizing grades in which only one element was varied in the initial part of the work. The multiplying factors were tested subsequently in multi-element steels and modified empirically to provide more widely applicable, averaged factors. However, the authors were unable to develop a single factor for molybdenum due to interactive effects and therefore separate factors are shown for this element for use in low- and high-nickel steels.

Although the effects can vary significantly with the carbon content and base composition, a guide to the potency of elements in the general group in promoting hardenability is shown below:

Vanadium
Molybdenum
Chromium
Manganese   decreasing effect
Silicon
Copper
Nickel

However, carbon could be placed at the top of this list and elements such as phosphorus and nitrogen, although present in small amounts, appear to produce hardenability effects of a similar magnitude to carbon. Although vanadium has a powerful effect on hardenability, it has a low solubility in steels due to the formation of vanadium carbide. Therefore the level of addition is generally small and, as illustrated shortly, may be used more to retard the tempering process than as a hardenability agent.

Boron

In the context of low-alloy engineering steels, boron is a unique alloying element from the following standpoints:

1. The addition of 0.002–0.003% B to a suitably protected base composition produces a hardenability effect comparable with that obtained from 0.5% Mo, 0.7% Cr or 1.0% Ni.
2. The effect of boron on hardenability is relatively constant provided a minimum level of soluble boron is present in the steel.
3. The potency of boron is related to the carbon content of the steel, being very effective at low carbon contents but decreasing to zero at the eutectoid carbon level.

Because of its high affinity for oxygen and nitrogen, boron is added to steel in conjunction with even stronger oxide- and nitride-forming elements in order to produce metallurgically active, soluble boron. In electric arc steelmaking, this involves additions of about 0.03% Al and 0.03% Ti, either separately or in the form of proprietary compounds containing the required levels of boron, aluminium and titanium. Without these additions, boron would react with oxygen and nitrogen in the steel and form insoluble boron compounds which have no effect on hardenability.

Gladman[8] has shown that the location of boron in steels is very dependent on the heat treatments that are applied. In low-alloy steels, boron was distributed uniformly throughout the microstructure of a 25.4 mm bar after water quenching from the austenitic range. However, in air-cooled samples, the grain boundaries were enriched in boron compared with the body of the grains. Ueno and Inoue[9] also investigated the presence of boron in a 0.1% C 3.0% Mn steel and showed that boron first segregated to and then precipitated at the grain boundaries, according to a typical 'C' curve pattern. They also showed that an increase in boron content decreased the incubation periods for segregation and precipitation, and in a steel containing 0.002% B, solution treated at 1350°C, quenching in iced brine was required in order to prevent the segregation of boron to the grain boundaries. It would appear therefore that under normal heat treatment conditions, involving oil quenching from temperatures of 820–920°C, boron segregates to the austenite grain boundaries and suppresses the formation of high-temperature transformation products.

Llewellyn and Cook[6] carried out a detailed investigation of the metallurgy of boron-treated engineering steels containing a wide range of carbon contents. The effect of boron content on hardenability was studied in a base composition of 0.2% C 0.5% Ni 0.5% Cr 0.2% Mo (SAE 8620) and the following Jominy hardenability criteria were examined:

1. Jominy distance to [hardness at J 1.25 mm – 25 HV].
2. Jominy distance to 350 HV, i.e. near the inflexion in the hardenability curve.
3. Hardness at J 9.8 mm, equivalent to the cooling rate at the centre of an oil-quenched, 28 mm bar.

As illustrated in Figure 4.15, each of these criteria reaches a maximum at a soluble boron content of about 0.0007%. Further additions produce a reduction

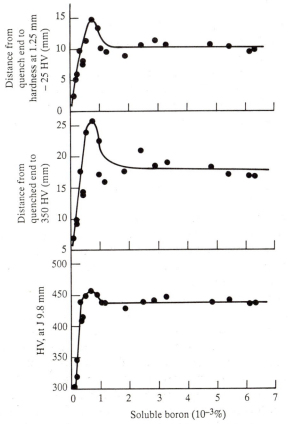

Figure 4.15 *Effect of boron on hardenability (After Llewellyn and Cook[6])*

in hardenability but a reasonably steady-state condition is achieved at boron contents in excess of 0.0015%. A similar pattern of results was also observed by Kapadia *et al.*[10] and, in commercial practice, it is usual to aim for soluble boron contents of 0.002–0.003%, accepting a slight loss in hardenability in favour of a consistent hardenability effect.

The interaction between boron and carbon was also investigated in SAE 8600 base steels and the Jominy hardenability curves for boron-free and boron-bearing versions of these steels are shown in Figures 4.13 and 4.16. These figures also indicate the formation of retained austenite at carbon contents in excess of 0.8%. However, the interaction between boron and carbon is illustrated in Figure 4.17 where the hardenability criteria examined are the Jominy distances

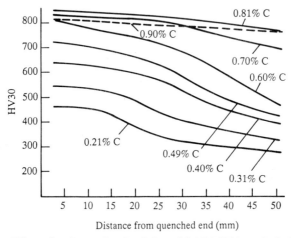

Figure 4.16   *Effect of carbon on hardenability of SAE 86B00 steels (After Llewellyn and Cook[6])*

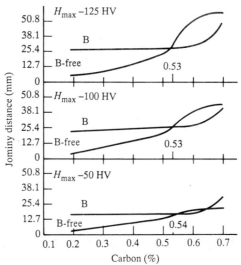

Figure 4.17   *Effect of boron on hardenability at various carbon levels (After Llewellyn and Cook[6])*

which coincide with decreases in hardness of 50, 100 and 125 HV below that obtained at the quenched end of the specimen (H_{max}). This figure shows that boron produces a marked increase in hardenability at a carbon content of 0.2% but the effect is steadily reduced to zero as the carbon content is increased to a level of 0.53–0.54%. Above this carbon level, boron has a detrimental effect on hardenability.

The above type of experiment was also carried out in base steels containing 0.8% Mn and the results are summarized in Figure 4.18. In this case, the efficiency of boron as a hardenability agent is expressed by means of a multiplying factor BF, calculated on the basis of:

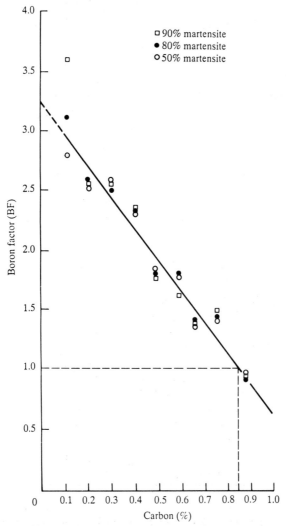

Figure 4.18 *Effect of carbon on boron multiplying factor (After Llewellyn and Cook[6])*

$$BF = \frac{\text{hardenability of (base steel + boron)}}{\text{hardenability of base steel}}$$

Thus the effect of boron on hardenability is again reduced with increase in carbon content but, in this particular base composition (0.8% Mn), reaches a value of zero (BF = 1.0) at a carbon content of about 0.85%.

From the above data, it is evident that the boron effect varies not only with carbon content but also with the alloy content of the base steel. Llewellyn and Cook have proposed that the critical carbon content may well correspond to the eutectoid level for the compositions concerned and that bainite might be promoted at the expense of martensite in boron-treated, hyper-eutectoid steels.

A consequence of the above effect is that the case hardenability of a carburized boron-treated steel is lower than that of a boron-free steel of comparable core hardenability. However, the limiting section sizes that will provide adequate case depths are probably well in excess of those that present a problem with regard to core hardenability.

Tempering resistance

Although the objective of quenching a steel is generally to produce a martensitic structure of high strength, steels are rarely put into service in the as-quenched condition because this represents a state of high stress, low toughness and poor ductility. After quenching, components are therefore tempered at an elevated temperature in order to obtain a better combination of properties. In carburized gears, the tempering treatment might be carried out at a temperature as low as 180°C, the purpose being to achieve relief of internal stresses without causing any significant softening in either the case or core of the components. On the other hand, medium-carbon steels containing about 0.4% C might be tempered at temperatures up to 650°C which results in a significant decrease in strength compared with the as-quenched condition but which is necessary in order to produce adequate toughness and ductility.

The mechanism of tempering involves the partial degeneration of martensite via the diffusion of carbon atoms out of solid solution to form fine carbides. If the tempering treatments are carried out at sufficiently high temperatures for long periods, then the breakdown of martensite can be complete, the micro-structure consisting of spheroidized carbide in a matrix of ferrite. However, treatments that are carried out to achieve this type of microstructure would be termed *annealing* rather than tempering.

Alloying elements affect the tempering process by retarding or suppressing the formation of Fe_3C, either by stabilizing the ε-carbide ($Fe_{2.4}C$) which is formed initially in the breakdown of martensite or by forming carbides that are more stable and grow more slowly than Fe_3C. Alloying elements therefore provide the opportunity of tempering at higher temperatures in order to obtain a higher ductility for a given strength level. Alternatively, improved tempering resistance might be employed to allow a component to operate at a higher temperature without softening.

Smallman[11] has produced the information given in Table 4.1 on the effect of alloying elements on tempering. In this table, the negative value ascribed to carbon indicates an acceleration in the tempering process which is due presumably to the increased supersaturation/driving force effect.

Table 4.1

Element	Retardation in tempering per 1% addition
C	− 40
Co	8
Cr	0
Mn	8
Mo	17
Ni	8
Si	20
V	30
W	10

Work by Grange and Baughman[12] established the following rank order of potency in promoting tempering resistance:

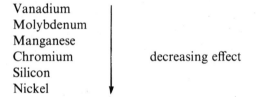

Vanadium
Molybdenum
Manganese
Chromium   decreasing effect
Silicon
Nickel

Thus there is broad agreement between the two sets of data, although Smallman indicates a relatively higher effect for silicon.

Although alloying elements such as vanadium and molybdenum are effective in promoting both hardenability and tempering resistance, they are expensive and the design of composition in engineering steels is dictated as much by costs as property requirements. Thus hardenability is achieved most cheaply through additions of manganese, chromium and boron but with little contribution to tempering resistance. However, such a condition may be perfectly acceptable in a case-carburized component where the tempering treatment is carried out at a low temperature. On the other hand, when the property requirements or operating conditions dictate the use of high tempering temperatures, resort is made to molybdenum and vanadium additions, but at relatively modest levels. At levels greater than about 1%, molybdenum induces a *secondary hardening* reaction.

In addition to its effect on tempering resistance, molybdenum is also added to steels to suppress temper embrittlement. This involves the co-segregation of alloying elements such as manganese and silicon and impurity elements such as antimony, arsenic, tin and phosphorus to the prior austenite grain boundaries in steels with bainitic or martensitic microstructures. Temper embrittlement occurs during exposure in the temperature range 325–575°C, either from operating within this range or by slow cooling through the range from a higher tempering temperature. This topic will be discussed further when dealing with *Steels for steam power turbines*, later in this chapter.

Whereas nickel is a well known alloying element in the context of engineering steels, it is expensive and not very effective in enhancing either hardenability or tempering resistance. However, it is perceived to be a ferrite strengthener and is used in engineering steels in order to promote toughness.

Surface-hardening treatments

Case carburizing

Case carburizing involves the diffusion of carbon into the surface layers of a low-carbon steel by heating it in contact with a carbonaceous material. The objective is to produce a hard, wear-resistant case with a high resistance to both bending and contact fatigue, whilst still maintaining the toughness and ductility of the low-carbon *core*. Carburizing is carried out at temperatures in the range 825–925°C in solid, liquid or gaseous media but, in each treatment, the transport of carbon from the carburizing medium takes place via the gaseous state, usually CO.

Carburizing in a solid, granular medium, such as charcoal, is termed *pack carburizing* and has been practised since ancient times. However, whereas the early treatments relied solely on the reaction between charcoal and atmospheric oxygen for the generation of CO_2 and CO, *energizers* such as sodium or barium carbonate are now added to the carburizing compound. During the heating-up period, the energizer breaks down to form CO_2 which then reacts with carbon in the charcoal to form CO:

$$BaCO_3 \rightarrow BaO + CO_2 \qquad (1)$$

$$CO_2 + C \rightarrow 2CO \qquad (2)$$

In turn, the CO reacts at the steel surface to form atomic carbon which defuses rapidly into the austenitic structure:

$$2CO \rightarrow C + CO_2 \qquad (3)$$

The CO_2 produced in this reaction then reacts with charcoal, reproducing reaction (2), and the cycle is repeated.

Pack carburizing is generally carried out at a temperature of 925°C and a case depth of about 1.5 mm can be obtained after carburizing for eight hours at this temperature. After this treatment, the component is removed from the carburizing compound and heat treated by various forms of quenching which will be described later.

Carburizing in liquid media generally takes place in molten salts (*salt bath carburizing*) in which the active constituent is sodium cyanide (NaCN), potassium cyanide (KCN) or calcium cyanide $Ca(CN)_2$. Oxygen is available at the salt bath–atmosphere interface and carburizing again takes place via the generation of CO. However, nitrogen is also liberated from the cyanides and diffuses into the steel. The amount of carbon and nitrogen absorbed by the steel is related to the temperature and cyanide content of the bath. With a NaCN content of 50%, the surface concentrations of carbon and nitrogen are of the order of 0.9% and 0.2% respectively, after treatments involving $2\frac{1}{2}$ hours at 900°C. However, salt bath carburizing is mainly used for small parts and treatments for $\frac{1}{2}$–1 hour produce a case depth of about 0.25 mm.

Gas carburizing is carried out in hydrocarbon gases such as propane (C_3H_8) or butane (C_4H_{10}) in sealed furnaces at temperatures of about 925°C. Major strides have been made in the control technology of the process and, since the 1950s, gas carburizing has become the most important method of case hardening. During the heating-up stage, the component is surrounded by an inert or reducing atmosphere. This atmosphere is referred to as the *carrier gas*, which is commonly an endothermic gas but may also be nitrogen based. On reaching the carburizing temperature, the furnace atmosphere is enriched to the required carbon level (*carbon potential*) by the addition of hydrocarbons which generate CO and carbon is absorbed at the surface of the steel. The carbon potential is controlled by varying the ratio of hydrocarbon to carrier gas but a surface carbon content of 0.8–0.9% is generally employed. A four-hour treatment at 925°C will produce a case depth of about 1.25 mm.

Following the carburizing operation, the components are subjected to hardening heat treatments, involving different forms of quenching:

- *Direct quenching* – quenching directly from the carburizing temperature
- *Single quenching* – allowing the component to cool to a temperature of about 840°C from the carburizing temperature before quenching
- *Reheat quenching* – cooling to room temperature and then reheating to a temperature of about 840°C before quenching
- *Double reheat quenching* – reheating first to about 900°C and quenching to produce a fine-grained core, followed by a second quenching treatment from a lower temperature, such as 800°C, in order to produce a fine-grained case region.

In case carburizing, single quenching is generally preferred to direct quenching because it reduces the thermal gradients in the steel, thereby minimizing dimensional movement or *distortion*. This topic will be discussed in the next section. Reheat quenching is employed in pack carburizing and also in gas carburizing if machining operations must be carried out before the final hardening treatment. Double reheat quenching is now virtually obsolete since satisfactory grain refinement can now be obtained in the core and case of modern fine-grained steels in the shorter quenching practices.

The type of quenching medium employed, e.g. oil or polymer, depends on the mechanical properties required but, as illustrated shortly, the quenching rate may have to be controlled very carefully in order to minimize dimensional movement. Following the quenching operation, carburized components are generally tempered in the range 150–200°C for periods of 2–10 hours in order to produce some stress relief in the high-carbon martensitic case.

Carburizing is used extensively in the automotive industry for the treatment of shafts and gears. It is also an important process in the production of large bearings in which the required level of hardness and fatigue resistance cannot be achieved in through-hardening grades such as 1.0% C 1.5% Cr (SAE 52100).

Nitriding

As its name suggests, *nitriding* involves the introduction of nitrogen into the surface of a steel but, unlike carburizing, it is carried out in the ferritic state at temperatures of the order of 500–575°C. However, like carburizing, it can be performed in solid, liquid or gaseous media but the most common is that involving ammonia gas (*gas nitriding*) which dissociates to form nitrogen and hydrogen:

$$2NH_3 \rightarrow 2N + 3H_2$$

Nascent, atomic nitrogen diffuses into the steel, forming nitrides in the surface region.

In *salt bath nitriding*, mixtures of NaCN and KCN are employed and the holding times are rarely longer than two hours, compared to periods of 10–100 hours in gas nitriding. A variation of salt bath nitriding is the *Sulfinuz* process which involves the addition of sodium sulphide (Na_2S) to the bath. This results in the absorption of sulphur into the steel as well as the introduction of carbon and nitrogen which is characteristic of conventional salt bath nitriding. The presence of sulphur in the surface of the component improves the anti-frictional behaviour and also the corrosion resistance of the steel.

Nitriding is carried out on steels containing strong nitride-forming elements such as Al, Cr and V and in BS 970 (*Wrought steels for mechanical and allied purposes*), the grades of steel shown in the Table 4.2 are identified specifically as being suitable for nitriding.

Table 4.2

Grade	C%	Si%	Mn%	Cr%	Mo%	V%	Al%
709M40	0.4	0.25	0.85	1.05	0.3		
722M24	0.24	0.25	0.55	3.25	0.55		
897M39	0.39	0.25	0.55	3.25	0.95	0.2	
905M39	0.39	0.25	0.55	1.6	0.2		1.1

Given the low temperatures involved in the process, nitriding can be carried out after the conventional hardening and tempering treatments have been applied to through-hardened steels.

Carbonitriding (nitro-carburizing)

Carbonitriding can be regarded as a variant of gas carburizing in which both carbon and nitrogen are introduced into the steel surface. This is achieved by the introduction of ammonia gas into the carburizing atmosphere which cracks, liberating nascent nitrogen. In some respects, the term *carbonitriding* may be misleading in that it is completely different from the nitriding process that takes place at low temperatures in the ferritic state and the term *nitro-carburizing* might be more appropriate. However, the process is carried out at temperatures of the order of 870°C and the case depths are lower than those produced by gas carburizing, e.g. a four-hour carbonitriding treatment at this temperature produces a case depth of about 0.75 mm.

The introduction of nitrogen produces a significant increase in the hardenability of the case region such that high surface hardness levels can be produced in steels of relatively low alloy content.

Induction hardening

In the processes described above, the hardness of the surface is increased by modifying the chemical composition of this region. In *induction hardening*, the composition of the material is unchanged but the surface is hardened by selective heat treatment. This is achieved by induction heating but a less controlled effect can be produced by the direct impingement of an oxy-acetylene torch in the process of *flame hardening*.

This process is generally applied to steels containing 0.30–0.50% C which give hardness values in the range 50–60 HRC. The steels may be C–Mn or low-alloy grades and induction hardening is carried out in the normalized or quenched and tempered condition, depending upon the section size or the properties required in the core.

Relative merits of surface treatments

In presenting a detailed review of gas carburizing, Parrish and Harper[13] examined the benefits of this process in relation to other surface-hardening treatments such as nitriding and induction hardening. Their main findings can be summarized as follows:

- *Carburizing* – capable of producing a wide range of case depths and core strengths and providing good resistance to bending and contact fatigue. The main drawback of the process is the distortion that occurs due to the thermal gradients induced by quenching from the austenitic range.
- *Nitriding* – produces relatively shallow hardening, e.g. a highly alloyed grade such as 897M39 ($3\frac{1}{4}$% Cr–Mo–V) has an effective case depth (at 500 HV) of only 0.35 mm after nitriding for 80 hours. However, nitriding is a distortion-free process that produces a surface which is resistant to scuffing and adhesive wear.
- *Induction hardening* – capable of producing a wide range of case depths using a range of compositions. Capable of producing similar contact fatigue and wear resistance to case carburizing and the process produces little distortion.

Distortion in case-carburized components

When transformable steels are heat treated, the volume changes that occur during heating and cooling (quenching) are not completely complementary and a component will exhibit a small change in shape compared with its original, pre-heat-treated condition. The term *distortion* is widely used to describe such changes in shape and represents a significant problem in the production of precision engineering components such as automotive gears. In such components, slight inaccuracies in shape lead to irregular tooth contact patterns which can result in problems ranging from a high level of noise in a gearbox or back axle to an overload situation which produces premature fatigue failure. The effect is therefore very important commercially but, given that dimensional change is inevitable under fast-cooling conditions, the approach to the problem is one of control and consistency of response rather than elimination.

In the mid-1960s, Murray[14] published work on the effects of composition on distortion in carburizing steels, using the *Navy C* specimen. This consisted of a split ring in which the dimensional change is measured by the degree of gap opening after heat treatment. Whereas some interesting results were obtained in both the UK and United States with this specimen, it had two major limitations:

1. The unrestrained nature of the split ring could result in gap openings of up to 1.2 mm which were very much larger than the dimensional changes that occur in automotive gears.
2. The effective section size of the specimen was small which made it unsuitable for the investigation of steels with medium to high hardenability.

Llewellyn and Cook[15] therefore designed a new specimen for the investigation of distortion which was washer-shaped with an outer diameter (OD) of 132 mm, bore diameter (BD) of 44 mm and a thickness (T) of 22 mm. These dimensions represented a compromise between the section size that was considered typical of a medium-size truck gear and the limiting size of specimen that could be produced from 50 kg experimental casts of steel. The specimens were carburized at 925°C for $5\frac{1}{2}$ hours at a carbon potential of 0.8%, furnace cooled to 840°C and then cooled at various rates to room temperature.

It was shown that hardenability has a marked effect on dimensional change, as illustrated in experimental casts of steel containing 1–3% Cr. The Jominy hardenability curves for these steels are shown in Figure 4.19 and the dimensional changes that occurred on oil quenching are illustrated in Figure 4.20. As the hardenability of the steels is increased, progressive contractions occur in the OD and BD and these are compensated by increases in thickness. This pattern of results was also repeated in other series of steels of different alloy content. As the alloy content and hardenability are increased, the transformation temperatures are depressed from those involving upper bainite to lower bainite and, ultimately, martensite. With regard to distortion, the consequence of this progression is the increasing volume change on transformation. Therefore, the progressively greater movement that results from increasing hardenability is associated with the progressive increase in volume on transformation. However,

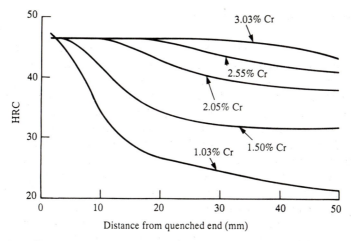

Figure 4.19 *Effect of chromium on Jominy hardenability (After Llewellyn and Cook[15])*

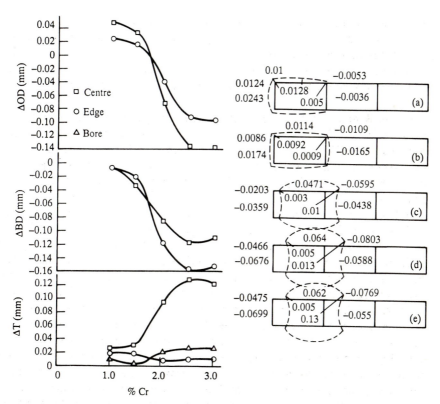

Figure 4.20 *Effect of chromium on dimensional change. Left top: outer diameter change. Left centre: bore diameter change. Left bottom: thickness change. Right: (a) 1.03% Cr; (b) 1.5% Cr; (c) 2.05% Cr; (d) 2.55% Cr; (e) 3.03% Cr (After Llewellyn and Cook[15])*

such an explanation does not address the reason for the irregular changes in shape that occur in these specimens but these are obviously associated with the complex stress situation that arises from the differential cooling rates that take place in a disc or component of limited symmetry.

One departure from the general trend between hardenability and distortion was observed in the case of boron-treated steels which behaved in an entirely different manner from boron-free steels of comparable hardenability. A satis-factory explanation for the unique behaviour of boron steels has not been proposed but they differ from boron-free steels in the following respects:

1. The case hardenability of a boron steel is lower than that of a boron-free steel of the same core hardenability.
2. Boron-treated steels will tend to have a higher M_s–M_f transformation range than boron-free steels of comparable hardenability.

Llewellyn and Cook showed that carburizing reduces the amount of movement that occurs in low-alloy/low-hardenability steels and related the effect to the depression of the M_s–M_f transformation range in the case-carburized region of a component. It was observed that carburizing had a smaller effect on dimen-sional change in highly alloyed steels and this may be related to the fact that these steels have low transformation ranges.

Cooling rate from the austenitic range was shown to have a dramatic effect on dimensional movement and water quenching can virtually eliminate the influence of other important factors such as carburizing treatment and alloy content/hardenability. This was illustrated in samples of commercial steels which were air cooled, slow oil quenched, fast oil quenched and water quenched, following a carburizing treatment at 925°C and furnace cooling to 840°C. The compositions of these steels are given in Table 4.3 and the hardenability curves are shown in Figure 4.21. The mean dimensional changes that took place after heat treatment are illustrated in Figure 4.22. In this figure,

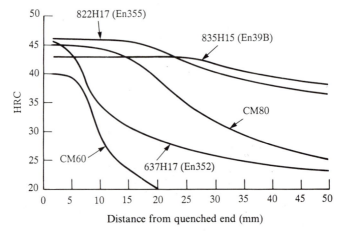

Figure 4.21 *Jominy hardenability of commercial steels (After Llewellyn and Cook[15])*

Table 4.3 Steels used to investigate the effect of quenching rate on distortion

Grade	C	Si	Mn	P	S	Cr	Mo	Ni	Al	B	Ti
637H17 (En 352)	0.18	0.28	0.81	0.011	0.037	0.92	0.1	1.1	:	:	:
822H17 (En 355)	0.18	0.34	0.6	0.025	0.016	1.47	0.2	2	0.056	:	:
835H15 (En 39B)	0.14	0.35	0.42	0.012	0.018	1.13	0.21	3.9	0.03	:	:
CM60	0.14	0.26	0.98	0.008	0.049	0.31	0.12	0.15	0.014	0.003	ND[a]
CM80	0.19	0.35	1.37	0.009	0.045	0.28	0.12	0.17	0.056	0.003	0.051

[a]ND = not determined.

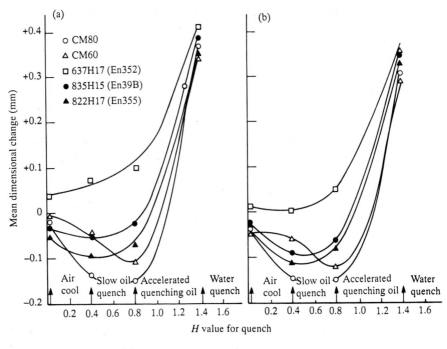

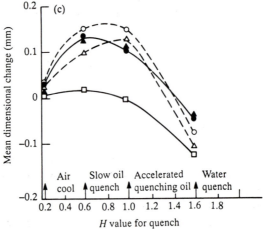

Figure 4.22   *Effect of quenching rate on mean dimensional change in (a) outer diameter, (b) bore diameter and (c) middle thickness (After Llewellyn and Cook[15])*

the various quenching treatments are shown in terms of the H value, an expression of the severity of the quenching rate. After air cooling, relatively little movement occurs in any of the steels, but as the quenching rate is increased to an H value of 0.8, most of the steels exhibit a significant decrease in OD and BD and a complementary increase in thickness. However, when the specimens are water quenched, a marked increase takes place in OD and BD and, despite the marked variation in hardenability, each of the steels tends to exhibit similar levels of movement.

From the foregoing remarks, it can be appreciated that a number of measures can be taken to reduce or control the degree of distortion that occurs in carburized components:

1. In some cases, quenching presses can be used to restrain the amount of movement that occurs.
2. Given the marked effect of hardenability, automotive manufacturers in particular will generally order steels to restricted hardenability bands, i.e. to only a half or even a third of the band width of the standard specification. This will minimize the variation in distortion such that the changes can be predicted and accommodated in the preheat treatment geometry.
3. The major effect of cooling rate is widely recognized in the automotive industry. Special quenching oils or warm oils are sometimes used in order to reduce temperature gradients during quenching, thereby reducing the degree of irregular dimensional change.

Standard specifications

The major UK standard for engineering steels is BS 970: Part 1: 1991 *Wrought steels for mechanical and allied engineering purposes*. It deals with steels of the following type in the form of blooms, billets, slabs, bars, rods and forgings:

- As-rolled, as-rolled and softened, and micro-alloyed C and CMn steels (Section 2).
- Through hardening boron steels (Section 3).
- Case hardening steels (Section 4).
- Stainless and heat resisting steels (Section 5).

Acknowledging that the standard contains a large number of grades, this edition of BS 970 has continued the practice of separating the steels into two categories. Category 1 steels are the recommended series of steels for use in new designs and in established designs whenever possible and are printed throughout the standard in normal (upright) type. Category 2 steels make up the remainder and are shown in *italic* (sloping) type.

The old En numbering system which became established after the Second World War was withdrawn some years ago and replaced by a logical, if complex, system which is outlined below.

In all designations, apart from those for stainless steels, the last two digits indicate the mean of the specified carbon content.

Each designation carries a letter which defines the supply requirements:

- M – for steels to specified mechanical properties
- H – for supply to hardenability requirements
- A – when supplied only to a specified analysis

In carbon and carbon-manganese steels, the first three digits indicate the mean of the manganese content and examples are shown in Table 4.4.

Table 4.4

Grade	C%	Mn%	Supplied to
060A62	0.6–0.65	0.5–0.7	Analysis only
080H46	0.43–0.5	0.6–1	Hardenability requirements
150M36	0.32–0.4	1.3–1.7	Mechanical properties

Free-cutting steels are now included in BS 970: Part 3: 1991 *Bright bars for general engineering purposes*. In these grades, the first three digits are in the range 200 to 240 and the second and third digits indicate the minimum or mean of the sulphur content multiplied by 100 (Table 4.5).

Table 4.5

Grade	C%	Mn%	S%	Supplied to
226M44	0.4–0.48	1.3–1.7	0.22–0.3	Mechanical properties

In the case of stainless steels, the designations reflect the well established AISI system. Thus the first three digits are in the 300 or 400 ranges, representing the austenitic and martensitic/ferritic grades respectively. The intervening letter S also signifies a stainless steel but the last two digits, chosen arbitrarily in the range 11 to 99, indicate variants within the main type. Examples are shown in Table 4.6.

Table 4.6

Grade	C%	Cr%	Ni%
420S29	0.14–0.2	11.5–13.5	–
420S37	0.2–0.28	12–14	–
431S29	0.12–0.2	15–18	2–3
304S15	0.06 max.	17.5–19	8–11
310S31	0.15 max.	24–26	19–22

Quenched and tempered grades are no longer included in BS 970, following the publication of the following European standards:

BS EN 10083-1: 1991   *Quenched and tempered steels (special steels)*
BS EN 10083-2: 1991   *Quenched and tempered steels (unalloyed quality steels)*

In the former, a useful comparison is provided of grades in the new European standard and equivalent grades in previous national standards. This information is shown in Table 4.7.

Table 4.7 *Comparison of steel grades specified in BS EN 10083-1: 1991, ISO 683-1 and other steel grades previously standardized nationally*

EN 10083-1	ISO 683-1: 1987[1]	Germany[1]		Finland	United Kingdom[1]	France[1]	Sweden SS steel	Spain	
		Alpha-numeric Name	Material number					Name	Number
2 C 22	–	(Ck 22)	(1.1151)	–	(070M20)	[XC 18]	–	–	–
3 C 22	–	(Cm 22)	(1.1149)	–	–	[XC 18 u]	–	–	–
2 C 25	(C 25 E4)	Ck 25	1.1158	–	(070M26)	[XC 25]	–	C25K	F1120
3 C 25	(C 25 M2)	Cm 25	1.1163	–	–	[XC 25U]	–	C25K-1	F1125(1)
2 C 30	(C 30 E4)	Ck 30	1.1178	–	(080M30)	[XC 32]	–	–	–
3 C 30	(C 30 M2)	Cm 30	1.1179	–	–	[XC 32 u]	–	–	–
2 C 35	(C 35 E4)	Ck 35	1.1181	C 35	(080M36)	[XC 38 H 1]	1572	C35K	F1130
3 C 35	(C 35 M2)	Cm 35	1.1180	–	–	[XC 38 H 1 u]	–	C35K-1	F1135(1)
2 C 40	(C 40 E4)	Ck 40	1.1186	–	(080M40)	[XC 42 H 1]	–	–	–
3 C 40	(C 40 M2)	Cm 40	1.1189	–	–	[XC 42 H 1 u]	–	–	–
2 C 45	(C 45 E4)	Ck 45	1.1191	C 45	(080M46)	[XC 48 H 1]	1672	C45K	F1140
3 C 45	(C 45 M2)	Cm 45	1.1201	–	–	[XC 48 H 1 u]	–	C45K-1	F1145(1)
2 C 50	(C 50 E4)	Ck 50	1.1206	–	(080M50)	–	1674	–	–
3 C 50	(C 50 M2)	Cm 50	1.1241	–	–	–	–	–	–
2 C 55	(C 55 E4)	Ck 55	1.1203	–	(070M55)	[XC 55 H 1]	–	C55K	F1150
3 C 55	(C 55 M2)	Cm 55	1.1209	–	–	[XC 55 H 1 u]	–	C55K-1	F1155(1)
2 C 60	(C 60 E4)	Ck 60	1.1221	–	(070M60)	–	–	–	–
3 C 60	(C 60 M2)	Cm 60	1.1223	–	–	–	–	–	–
28 Mn 6	(28 Mn 6)	28 Mn 6	1.1170	–	(150M19)	–	–	–	–
38 Cr 2	–	38 Cr 2	1.7003	–	–	(38 C 2)	–	–	–
38 CrS 2	–	38 CrS 2	1.7023	–	–	(38 C 2 u)	–	–	–
46 Cr 2	–	46 Cr 2	1.7006	–	–	–	–	–	–
46 Cr S 2	–	46 CrS 2	1.7025	–	–	–	–	–	–

Table 4.7 *Continued*

EN 10083-1	ISO 683-1: 1987[1]	Germany[1] Alpha-numeric Name	Germany[1] Material number	Finland	United Kingdom[1]	France[1]	Sweden SS steel	Spain Name	Spain Number
34 Cr 4	34 Cr 4	34 Cr 4	1.7033	–	(530M32)	(32 C 4)	–	–	–
34 CrS 4	34 CrS 4	34 CrS 4	1.7037	–	–	(32 C 4 u)	–	–	–
37 Cr 4	37 Cr 3	37 Cr 4	1.7034	–	(530M36)	(38 C 4)	–	38Cr4	F1201
37 CrS 4	37 CrS 4	37 CrS 4	1.7038	–	–	(38 C 4 u)	–	38Cr4-1	F1206(1)
41 Cr 4	41 Cr 4	41 Cr 4	1.7035	–	(530M40)	42 C 4	–	42Cr4	F1202
41 CrS 4	41 CrS 4	41 CrS 4	1.7039	–	–	42 C 4 u	2245	42Cr4-1	F1207(1)
25 CrMo 4	25 CrMo 4	25 CrMo 4	1.7218	25 CrMo 4	(708M25)	25 CD 4	2225	–	–
25 CrMoS 4	25 CrMoS 4	25 CrMoS 4	1.7213	–	–	25 CD 4 u	–	–	–
34 CrMo 4	34 CrMo 4	34 CrMo 4	1.7220	34 CrMo 4	(708M32)	(34 CD 4)	2234	–	–
34 CrMoS 4	34 CrMoS 4	34 CrMoS 4	1.7226	–	–	(34 CD 3 u)	–	–	–
42 CrMo 4	42 CrMo 4	42 CrMo 4	1.7225	42 CrMo 4	(708M40)	42 CD 4	2244	40CrMo4	F1252
42 CrMoS 4	42 CrMoS 4	42 CrMoS 4	1.7227	–	–	42 CD 4 u	–	40CrMo4-1	F1257(1)
50 CrMo 4	50 CrMo 4	50 CrMo 4	1.7228	–	(708M50)	–	–	–	–
36 CrNiMo 4	36 CrNiMo 4	36 CrNiMo 4	1.6511	–	(817M37)	–	–	–	–
34 CrNiMo 6	(36 CrNiMo 6)	(34 CrNiMo 6)	(1.6582)	34 CrNiMo 6	(817M40)	–	2541	–	–
30 CrNiMo 8	(31 CrNiMo 8)	30 CrNiMo 8	1.6580	–	[823M30]	30 CND 8	–	–	–
36 NiCrMo 16	–	–	–	–	[835M30]	35 NCD 16	–	–	–
51 CrV 4	[51 CrV 4]	50 Cr V 4	1.8159	–	[735A50]	(50 CV 4)	–	51CrV4	F1430

Note:
[1] If a steel grade is given in round brackets, this means that the chemical composition differs only slightly from EN 10083-1. If it is given in square brackets, this means that greater differences exist in the chemical composition compared with EN 10083-1. If there are no brackets around the steel grade, this means that there are practically no differences in the chemical composition compared with EN 10083-1.

After BS EN 10083-1: 1991.

For many years, BS 970 adopted a system whereby a single letter was used to specify a particular range of tensile strength in the quenched and tempered condition. This system is still widely used in the UK and the full range of letters and the associated tensile ranges are shown in Table 4.8.

Table 4.8

Reference symbol	Tensile strength (N/mm^2)
P	550–700
Q	625–775
R	700–850
S	775–925
T	850–1000
U	925–1075
V	1000–1150
W	1075–1225
X	1150–1300
Y	1225–1375
Z	1550 min.

Work is in hand to prepare European standards on stainless steels and boron steels and BS 970 will then be withdrawn completely.

Former versions of BS 970: Part 1 contained an appendix which provided a guide to the selection of Category 1 through hardening steels, based on tensile strength and limiting ruling section. This guide was prepared for applications where the most important criterion was the level of mechanical properties required in the finished part and with the aim of assisting in the selection of the most cost-effective grade for a given tensile strength and section size. Although many of the steel designations are now obsolete with the introduction of European standards, it was considered worthwhile to maintain this information in the present text and this is shown in Table 4.9. The first column in this table (*Heat treatment condition*) lists the various ranges of tensile strength according to the lettering system P–Z defined above. Along each row of the table, the ruling section increases and the various grades of steel that will satisfy the required combination of tensile strength and section size are identified. Thus a simple C-Mn steel (080M30) is recommended for an application calling for a tensile strength of 625-775 N/mm² (Q condition) in small section sizes. On the other hand, a 2.5% Ni-Cr-Mo steel (826M40) is recommended when a tensile strength of 1075-1225 N/mm² (W condition) is required in section sizes greater than 150 mm.

Steel prices

Because of the very large number of steel grades covered by BS 970, one can only be very selective in attempting to provide an introduction to the cost structure of engineering steels. Purely on an arbitrary basis, the selection was

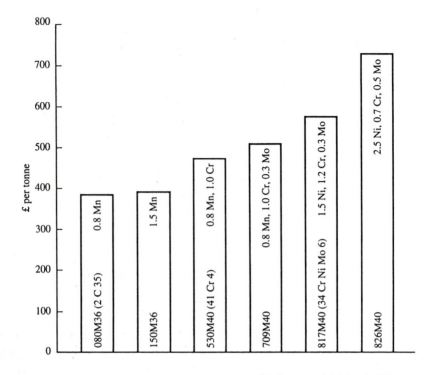

Figure 4.23   *Basis prices of engineering steel billets as of 1 March 1994*

narrowed down to some of the grades in the 19–29 mm ruling selection column in the Category 1 steels tabulation (Table 4.9). However, this selection was augmented by one further grade, namely 826M40, which is recommended for the attainment of the highest strength (1075–1225 N/mm^2) in the largest ruling section (150–250 mm) in this table.

The prices of the six steels concerned, in the form of large billets, are shown in Figure 4.23. This indicates a steady increase in price which reflects the progressive addition of alloying elements such as chromium, molybdenum and nickel.

Machinable steels

Machining is an important stage in the production of most engineering components and, in many automotive transmission parts, machining can account for up to 60% of the total production costs. It is not surprising therefore that the engineering industries have called for steels with improved and consistent levels of machinability, whilst still maintaining the other properties that ensure good service performance. The common machining processes include turning, milling, grinding, drilling and broaching and several of these operations might be carried out on an automatic lathe in the production of a single component. Each of these processes differs in terms of the metal cutting action and involves different conditions of temperature, strain rate and

Table 4.9 Guide to the selection of category 1 through hardening steels

Heat treatment condition	Tensile strength range (N/mm²)[a]	HB range	Ruling section						
			≤13 mm	>13 mm ≤19 mm	>19 mm ≤29 mm	>29 mm ≤63 mm	>63 mm ≤100 mm	>100 mm ≤150 mm	>150 mm ≤250 mm
Q	625–775	179–229	080M30	080M30	080M36	080M40, 150M19	120M36	135M44, 150M36, 080M50[b], 216M44[c]	708M40
R	700–850	201–255	212M36[c]	212M36[c]	120M36, 150M19, 080M46[b], 216M44[c], 226M44[c]	135M44, 150M36, 080M50[b], 216M44[c], 226M44[c]	135M44, 530M40, 070M55[b], 216M44[c], 606M36[c]	708M40, 605M36	708M40
S	775–925	223–277	120M36, 080M46[b], 216M44[c], 226M44[c]	120M36, 080M50[b], 216M44[c], 226M44[c]	135M44, 150M36, 080M50[b], 216M44[c], 226M44[c]	530M40, 070M55[b], 606M36[c]	708M40, 605M36	709M40	709M40

	Rm (N/mm²)	HB								
T	850–1000	248–302	135M44^d	080M50^{b,d} 216M44^{c,d} 226M44^{c,d}	135M44^d	530M40	708M40 605M36	709M40	817M40	817M40
U	925–1075	269–331	135M44^d	135M44^d	070M55^{b,d} 606M36^c	606M36^c	709M40	817M40	720M32	720M32
V	1000–1150	293–352	708M40 605M36	708M40 605M36	708M40 605M36	708M40 605M36	817M40	720M32	720M32	826M40
W	1075–1225	311–375	708M40^{d,e} 817M40^e	709M40^{d,e} 817M40^e	709M40^{d,e} 817M40^e	817M40^c	720M32^c	720M32^c	826M40^c	826M40^c

^{a}1 N/mm² = 1 MPa.
^{b}No specified impact properties.
^{c}Free cutting steel.
^{d}Full mechanical properties may not always be obtained by bulk heat treatment but the properties can be achieved by the appropriate heat treatment of die forgings and components.
^{e}Often ordered in the softened condition for machining and subsequent heat treatment.

After BS 970: Part 1: 1983

chip formation. Therefore the machining performance of a steel cannot be defined by means of a single parameter. However, some of the features that are often invoked as measures of machinability are:

- Tool life
- Production rate (e.g. components/hour)
- Power consumption
- Surface finish of component
- Chip form
- Ease of swarf removal

Following a brief description of machinability testing, the role of free cutting additives will be discussed. Attention will then be turned to the composition and machining characteristics of:

1. Low-carbon free cutting steels.
2. Medium-carbon steels.
3. Low-alloy steels.
4. Stainless steels.

Machinability testing

Many organizations have developed their own individual test methods for evaluating machinability and these can range from tests involving tool failure in only a few minutes of metal cutting to those simulating commercial practices and lasting several hours. However, reproducibility is of absolute importance and, regardless of the type of test employed, the cutting conditions such as *speed* (e.g. peripheral bar speed in m/min), *feed* (rate of travel of cutting tool in mm/rev) and *depth of cut* (metal removed per cut in mm) must be carefully controlled.

A widely adopted laboratory test for machinability is the Taylor Tool Life test in which the life of the tool is determined at various cutting speeds. As illustrated in Figure 4.24, plots of tool life (T) against cutting speed (V) provide a straight line giving the equation:

$$VT^n = C$$

where C and n are constants.

One particular parameter of tool life which features prominently in machining evaluations is $V20$, the cutting speed that will provide a tool life of 20 minutes. However, such short-term tests may not necessarily provide an accurate guide to the performance in longer term industrial machining operations.

Although simple turning tests such as those described above are useful in steel development or for quality control purposes, steel users often call for longer term tests involving multi-machining operations. Thus up to two tonnes of bright drawn bar stock might be consumed over a period of about seven hours in tests involving the production of the type of test piece shown in Figure 4.25. This involves turning, drilling, plunge cutting and parting operations and the cutting rates are adjusted so as to maintain a specific dimensional tolerance in the component over a simulated shift period. The machinability rating of a

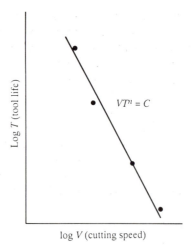

Log T (tool life)

$VT^n = C$

log V (cutting speed)

Figure 4.24   *Taylor tool life curve*

given steel sample can then be expressed in terms of components per hour or production time per component.

In each of the laboratory tests for machinability, rigorous attention must be given to maintaining standard characteristics in the cutting tools in terms of grade, hardness and tool geometry.

Role of free cutting additives

Various elements are added to steel in order to improve the machining performance. The main free cutting additives and their perceived action in improving machinability are as follows.

Sulphur

Sulphur is the cheapest and most widely used free cutting additive in steels. Whereas most specifications for engineering steels restrict the sulphur content to 0.05% max., levels up to about 0.35% S are incorporated in free cutting steels. Sufficient manganese is also added to such steels to ensure that all the sulphur is present as MnS rather than FeS, which causes *hot shortness* (cracking) during hot working. The MnS inclusions deform plastically during chip formation into planes of low strength which facilitate deformation in the primary shear zone.

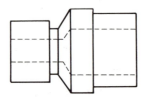

Figure 4.25   *Machinability test piece (By permission of British Steel Technical)*

The MnS inclusions also exude into the tool–chip interface, acting as a lubricant and also forming a protective deposit on the tool. The net effect is a reduction in cutting forces and temperatures and a substantial reduction in the tool wear rate. This is illustrated in Figure 4.26, which relates to low-carbon free cutting steels and shows the marked decrease in flank wear rate with increase in the volume fraction of MnS inclusions. Although relatively little improvement in wear rate is achieved at sulphide volume fractions greater than 1.5%, the chip form and surface finish continue to improve with further additions of sulphur.

For many years, it has been postulated that the morphology of the MnS inclusions plays a major role in the machining performance of free cutting steels, large globular inclusions being far more effective than thin elongated inclusions. However, sulphide morphology is influenced markedly by the state of deoxidation of the steel, a heavily killed steel promoting the formation of MnS inclusions that are easily deformed into elongated inclusions during hot working. In addition, killed steels tend to contain hard, abrasive oxide inclusions which have an adverse effect on machinability. Therefore it is extremely difficult to differentiate between the effects that might be due to sulphide morphology and those that are clearly due to the presence of hard, abrasive oxide inclusions.

One of the disadvantages of high sulphur contents is that they impair the transverse ductility of steels, particularly when MnS is present as long elongated inclusions. As indicated later, inclusion-modifying agents can be added to high-sulphur steels in order to promote a more favourable sulphide morphology.

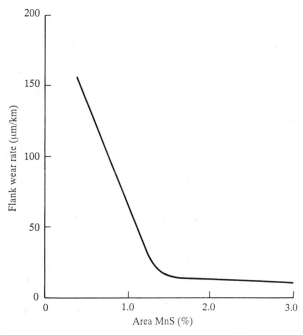

Figure 4.26 *Effect of sulphide content on machinability (After Wannel et al.[16])*

Lead

After sulphur, lead is the next most common additive and additions of 0.15–0.35% Pb are incorporated in free cutting steels. Such levels are soluble in molten steel but are precipitated as discrete particles of lead during solidification. Due to its high density, major precautions have to be taken in the production of leaded steels in order to avoid segregation effects. Again lead reduces the frictional effects at the tool–chip interface, where it becomes molten at the elevated temperatures generated during cutting. Lead is also thought to have an embrittling effect in the primary shear zone, thereby shortening the chips and improving surface finish. The lead particles are often present as tails to the MnS inclusions.

Lead has little effect on the mechanical properties of steel at ambient temperature since it is generally present as a globular constituent.

The toxic effects of lead are well known and care has to be taken to contain lead fumes during the production of leaded steels. Similar measures are also taken during the drop forging of leaded steels but the author is not aware that hygiene problems arise during the machining of these steels.

Tellurium

Tellurium is an efficient but relatively expensive addition in free cutting steels and is generally restricted to a maximum level of 0.1%. In larger amounts, it leads to cracking during hot working. Tellurium is generally present as manganese telluride, which is a low melting point compound and should therefore act in a similar way to lead. The high surface activity of tellurium is also considered important in its action as a free machining additive. Small additions of tellurium, e.g. 0.01%, are also added to engineering steels in order to produce more globular MnS inclusions and promote better transverse properties.

Like lead, tellurium presents hygiene problems during steelmaking (*garlic breath*) and requires the operation of fume extraction systems.

Selenium

Selenium additions of 0.05–0.1% are made to low-alloy steels, but in free machining stainless steels many specifications call for a minimum of 0.15% Se. Selenium is present as a mixed sulphide–selenide and, like tellurium, small additions of this element are also effective in promoting more globular sulphide inclusions.

Bismuth

Bismuth is closely related chemically to lead and its free cutting properties have been known for many years. However, commercial interest in bismuth has only been significant since about 1980, as greater anxieties have been voiced about the use of lead. Levels of up to 0.1% Bi are typical in free cutting grades and bismuth is present as tails to the MnS inclusions. Its action in promoting improved machining characteristics appears to be similar to that of lead.

Calcium

As indicated earlier, oxide inclusions are hard and abrasive and detract from the machinability of steels. This is particularly the case with alumina inclusions which are formed in engineering steels due to the practice of adding aluminium as a deoxidant or grain-refining element. However, the adverse effects of alumina can be reduced with the addition of calcium, which results in the formation of calcium aluminate. These inclusions soften during high-speed machining and form protective layers on the surface of carbide tools. Calcium is also effective in reducing the projected length of MnS inclusions, thereby improving the transverse properties of resulphurized steels.

Low-carbon free cutting steels

For some engineering components, the mechanical property requirements are minimal and by far the most important requirement is a high and consistent level of machinability. Hose couplings and automotive spark plug bodies are examples of such components which are mass produced at high machining rates. Sulphur contents in the range 0.25–0.35% are typical but, as illustrated in Figure 4.27, substantial improvements in machinability are obtained by the addition of lead to resulphurized steels. Where extremely high rates of machining are required, a low-carbon free cutting steel might be treated with sulphur, lead and bismuth, e.g. 0.25% S, 0.25% Pb, 0.08% Bi.

The hardness of low-carbon free cutting steels is important and the effect appears to be associated with the achievement of an optimum embrittling effect

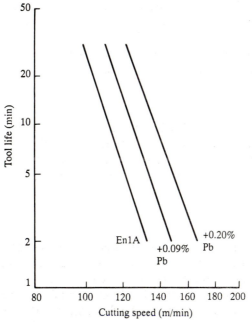

Figure 4.27   *Effect of lead on machinability (After Wannel et al.[16])*

for chip formation. This is illustrated in Figure 4.28, which indicates an optimum hardness of 105–110 HV in the normalized condition. Therefore the composition of these steels, including residual elements such as chromium, nickel and copper, needs to be controlled in order to achieve the optimum hardness. Additions of phosphorus and nitrogen may also be employed in achieving the desired embrittlement effects. Above the critical hardness level, embrittlement and ease of chip formation give way to increased abrasion and reduced tool life.

As indicated earlier, oxide inclusions are particularly damaging to machinability and the deoxidation practices adopted for low-carbon free cutting steels have to be controlled very carefully in order to minimize the level of these inclusions. With the move from ingot production to continuous casting, this problem has become more acute as the steels need to be more heavily deoxidized in order to minimize gas evolution during solidification.

Medium-carbon free cutting steels

Medium-carbon steels, containing 0.35–0.5% C and up to 1.5% Mn, are often used in the normalized condition for engineering components requiring a tensile strength of up to about 1000 N/mm². The free cutting versions of these steels are generally based on sulphur contents of 0.2–0.3% but, because of their higher strength, they are significantly harder to machine than low-carbon free cutting steels. Therefore benefit can be gained in selecting a grade of steel with the lowest level of carbon consistent with achieving the required strength in the end product.

Whereas silicate inclusions impair the machinability of low-carbon free cutting steels, they are not particularly damaging in medium-carbon steels when machined with carbide tools. This is due to the fact that higher temperatures are generated in the cutting of these steels, leading to the softening of the silicate

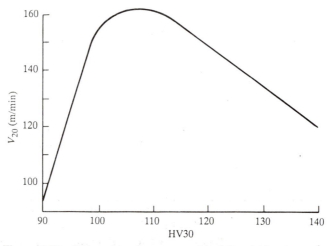

Figure 4.28   *Effect of hardness on machinability (After Wannel et al.[16])*

inclusions. However, alumina particles are again particularly detrimental, reducing the machining performance of both plain and resulphurized medium-carbon steels.

Machinable low-alloy steels

In high-strength automotive transmission components, a high level of toughness and ductility may be required in the transverse direction. For this reason, free machining versions of low-alloy engineering steels were based traditionally on the more costly lead additions rather than sulphur. However, as indicated earlier, the addition of inclusion-modifying agents can reduce the anisotropy in mechanical properties and provide improved service performance in resulphurized steels. This is illustrated in Figure 4.29, which shows the improvement in transverse impact properties obtained in SAE 4140 steel[17] with calcium treatment. Bearing in mind that high-strength engineering steels are generally fine grained with aluminium, calcium treatment will also lead to an improvement in machinability with carbide cutting tools due to the formation of calcium aluminate rather than alumina.

Machinable stainless steels

Although out of context in terms of chapter heading, it was thought that benefit might be gained in continuity of technology by dealing with machinable stainless steels at this stage.

Austenitic stainless steels have high rates of work hardening and this results in poor machinability. However, this problem has been exacerbated to some extent by modern steelmaking practices, such as arc-AOD, which produce low levels of sulphur, e.g. often less than 0.01%. To overcome this particular problem, the steels can be resulphurized to a level just below the maximum generally permitted in standard specifications, e.g. 0.03% max.

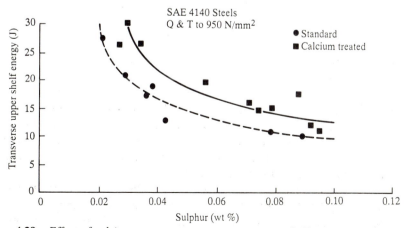

Figure 4.29 *Effect of calcium treatment on transverse upper shelf energy of SAE 4140 steels (After Pickett et al.[17])*

Like carbon and low-alloy steels, large additions of sulphur are also made to stainless steels in order to provide free cutting grades. Thus AISI 303 is an 18% Cr 9% Ni austenitic stainless steel containing 0.15% S min. However, the large volume fraction of MnS inclusions in this grade impairs the corrosion properties such that the performance is significantly worse than that of the standard 18% Cr 9% Ni (Type 304) stainless steel. Where this presents a significant problem, free cutting properties coupled with improved corrosion resistance can be obtained in Type 303Se, the 18% Cr 9% Ni grade containing 0.15% Se min.

Steels for gas containers

High-pressure gas cylinders have been in use since the 1870s when they were first introduced for the transportation of liquefied carbon dioxide for the aerated drinks industry. Since that time, their use has extended enormously to deal with the conveyance of a variety of *permanent gases*, such as air, argon, helium, hydrogen, nitrogen and oxygen, and other *liquefiable gases*, such as butane, propane, nitrous oxide and sulphur dioxide. Given the explosive, flammable or toxic nature of some of these gases, it is not surprising that the manufacture and utilization of gas containers are the subject of major scrutiny by the Home Office and also by international bodies. However, whilst still maintaining very high standards of safety, the use of higher strength steels has been permitted which has greatly increased the carrying efficiency of gas cylinders in terms of their gas capacity per weight.

In the UK, the relevant British Standard is BS 5045 *Transportable gas containers*, of which Parts 1 and 2 are the main sections for steel containers:

- Part 1:   Specification for seamless gas containers above 0.5 litre water capacity
- Part 2:   Steel containers up to 130 litre water capacity with welded seams

This topic is the subject of reviews by Irani[18] and Naylor.[19]

Steel compositions

BS 5045: Part 1: 1982 specifies four grades of steel for seamless cylinders, covering C, C–Mn, Cr–Mo and Ni–Cr–Mo compositions, and details are shown in Table 4.10. The most popular compositions are the C–Mn and Cr–Mo grades, which are hardened and tempered to provide a minimum yield strength value of 755 N/mm^2. However, as indicated later, the yield strength is restricted to lower levels for certain types of hydrogen containers.

In relation to welded containers, BS 5045: Part 2: 1989 permits the use of C or C–Mn steels but, as illustrated in Table 4.11, these grades provide minimum yield strength values in the range 215–350 N/mm^2, i.e. significantly lower than that attained in seamless containers.

Table 4.10 *Seamless steel gas containers to BS 5045: Part I: 1982*

Material	Code	C%	Si%	Mn%	Cr%	Mo%	Ni%	YS[a] (N/mm^2) min.	TS[a] (N/mm^2)
Carbon steels	M	0.15–0.25	0.05–0.35	0.4–0.9	–	–	–	250	430–510
	C	0.35–0.45	0.05–0.35	0.6–1	–	–	–	310	570–680
C–Mn steels	Mn	0.4 max.	0.1–0.35	1.3–1.7	–	–	–	445	650–760
	Mn H							755	890–1030
Cr–Mo alloy steel	CM	0.37 max.	0.1–0.35	0.4–0.9	0.8–1.2	0.15–0.25	0.5 max.	755	890–1030
Ni–Cr–Mo alloy steel	NCM	0.27–0.35	0.1–0.35	0.5–0.7	0.5–0.8	0.4–0.7	2.3–2.8	755	890–1030

[a] For containers for use in hydrogen trailer service, the minimum yield strength shall not exceed 680 N/mm² and the tensile strength shall be within the range 800–930 N/mm².

After BS 5045: Part I: 1982 Amendment No. 1, August 1986.

Table 4.11 *Welded steel gas containers to BS 5045: Part 2: 1989*

Chemical and physical properties	Type A	Type B	Type C	Type D	Type E	Type F	Type G
Carbon % max.	0.2	0.18	0.2	0.2	0.15	0.18	0.25
Silicon % max.	0.3	0.25	–	0.3	0.3	0.4	0.35
Manganese							
% min.	–	0.4	0.7	–	–	0.5	0.6
% max.	0.6	1.2	1.5	0.6	1.2	1.4	1.4
Phosphorus % max.	0.05	0.04	0.05	0.05	0.025	0.04	0.03
Sulphur % max.	0.05	0.04	0.05	0.05	0.03	0.04	0.045
Grain-refining							
elements % max.	–	–	–[a]	–	0.3[a]	–	0.7
Yield stress							
(N/mm^2) min.	215	275	310	200	350	285	250
Tensile strength							
(N/mm^2) min.	340	400	430	320	430	430	430
Tensile strength							
(N/mm^2) max.	430	490	585	420	650	510	550
Elongation % min.[b]							
$L_o = 50$ mm	28	24	21	29	21	–	21
$L_o = 5.65 \sqrt{S_o}$	33	29	20	35	25	20	25

[a] Grain-refining elements are limited to: niobium 0.08%, titanium 0.2%, vanadium 0.2%, niobium plus vanadium 0.2%.
[b] Where any other non-proportional gauge lengths are used, conversions are in accordance with BS 3894: Part 1.
Note 1. L_o is the original gauge length.
             S_o is the original cross-sectional area.
Note 2.   Type A and type C are equivalent to grades of BS1449: 1962 (withdrawn); type E is equivalent to grade 43/35 of BS 1449: Part 1; type D is similar to Euronorm 120; type G is equivalent to type 151 grade 430 of BS 1501: Part 1.
After BS 5045: Part 2: 1989.

Design and manufacture

The manufacture of a seamless steel cylinder is shown schematically in Figure 4.30, which illustrates the high degree of metal forming involved in this complex hot-working operation. The process is all the more remarkable when it is realized that the production of seamless cylinders from thick plate or billets was developed in the 1880s and 1890s and has changed very little to the present day.

The cylinder wall is the thinnest region of the vessel and is therefore subjected to the highest stresses. However, under fatigue conditions, the higher stresses are generated at the junction of the relatively thin wall and the thicker concave base. BS 5045: Part 1 details the following formulae for the calculation of minimum wall thickness in seamless containers:

$$t = \frac{0.3 \,(\text{test pressure}) \times (\text{internal diameter})}{7 \times f_e - (\text{test pressure})} \tag{1}$$

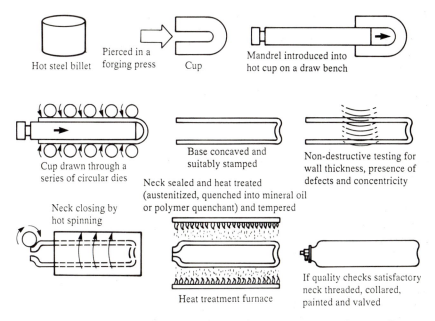

Figure 4.30   *The protection of seamless steel gas cylinders (After Irani[18])*

$$\text{or } t = \frac{0.3 \text{ (test pressure)} \times \text{(external diameter)}}{7 \times f_e - 0.4 \text{(test pressure)}} \qquad (2)$$

where f_e is the maximum permissible equivalent stress at test pressure.

In general, f_e is equal to $0.75 \times$ minimum yield strength. However, a value of $0.875 \times$ minimum yield strength is allowed in special portable containers, e.g. those used in aircraft, underwater breathing apparatus and life raft inflation, thereby facilitating the use of lighter containers. For a Cr–Mo steel cylinder (230 mm OD), used for the transportation of oxygen at 200 bar (equivalent to a test pressure of 300 bar), equation (2) provides a minimum wall thickness t of 5.39 mm.[18]

BS 5045: Part 2 provides the same equations for the calculation of minimum thickness as those shown earlier for seamless containers (equations (1) and (2)). However, as indicated previously, the proof strength values of steels specified in Part 2 are significantly lower than those included in Part 1. Thus for a given test pressure, welded containers would require a greater thickness than seamless containers but, in effect, the former operate at lower pressures.

Rather strangely, BS 5045 Parts 1 and 2 make no reference to impact test requirements although Oldfield[20] reports that *leak-before-burst* philosophies are growing in acceptance and that fracture toughness tests are being developed for use with gas containers. This is particularly pertinent when handling aggressive gases, as illustrated in the following section.

Hydrogen gas containers

During the 1970s and early 1980s, almost one hundred failures occurred in Europe (20 in the UK) in gas containers that were used for the delivery of hydrogen by road trailer. Failure was due to the propagation of fatigue cracks from small manufacturing defects which were present at the inner surface of the cylinder, close to the highly stressed junction where the wall joins the concave base. The fatigue cracks propagated through to the outside wall of the cylinder, resulting in leakage of gas. The failed containers had been made by various manufacturers and conformed to the appropriate specifications at that time. Under the aegis of the Health and Safety Executive, a committee was established to investigate the failure mechanism and make appropriate recommendations. The results of this investigation are the subject of an interesting paper by Harris *et al.*[21]

The problem containers in the UK were manufactured in Cr–Mo steel, which had been used for many years for the manufacture of storage containers without any serious problems. However, the transportable hydrogen containers differed from the storage containers in one important respect, namely the refilling frequency, which was the main source of fatigue cycling, the other being due to transportational stresses. Whereas the storage containers could experience up to 12 refills per annum, the transportable containers might be refilled twice daily. Harris *et al.* carried out fatigue crack growth tests on Cr–Mo and C–Mn steels in hydrogen gas at 150 bar and their results are shown in Figure 4.31. Whereas the bulk of the test data fell within a relatively narrow scatter band, the major exception was the Cr–Mo steel with the relatively high hardness of 324 HB. On the other hand, the fatigue crack growth of all the steels in hydrogen was very much greater than that induced in a nitrogen atmosphere. The acceleration factor by which the crack growth rate in hydrogen exceeds that in nitrogen at a stress intensity value of 20 $MNm^{-3/2}$ is shown as a function of hardness in Figure 4.32. This figure demonstrates very clearly the influence of the hardness (or strength) of the steel on the fatigue behaviour, the acceleration factor increasing from below 20 to above 100 as the hardness is increased from 274 to 324 HB.

Subsequent to the above work, amendments were incorporated into the standard for transportable gas containers and appear as Appendix E to BS 5045: Part 1 (Containers for use in hydrogen trailer service). This restricts the hardness range to 230–290 HB. In addition, the yield stress is not allowed to exceed 680 N/mm^2 and the tensile strength must be within the range 800–930 N/mm^2.

Higher strength steels

As indicated earlier, the carrying capacity of gas containers has been increased progressively over the years and this has been achieved primarily through the use of steels with progressively higher yield strength. This trend is continuing and Naylor[19] describes development work on gas container steels with minimum yield strength values of 950 N/mm^2. These higher strength values are achieved

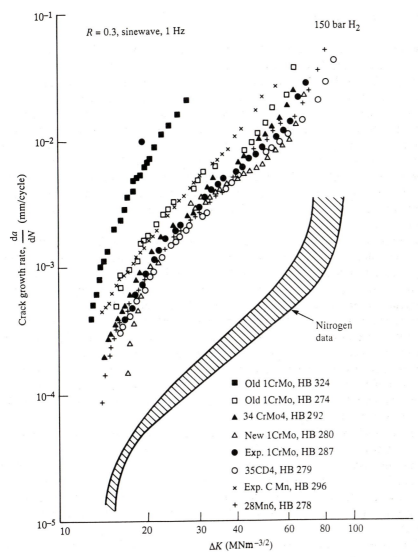

Figure 4.31 *The influence of hydrogen on the rate of fatigue crack growth in gas container steels (After Harris* et al.[21] *)*

with micro-alloying additions of vanadium and also with higher levels of molybdenum or silicon. Naylor reports that cylinder-manufacturing trials were being carried out on these experimental compositions.

Bearing steels

Bearings constitute vital components in most items of machinery, permitting accurate movement under low frictional conditions. In addition, they are also

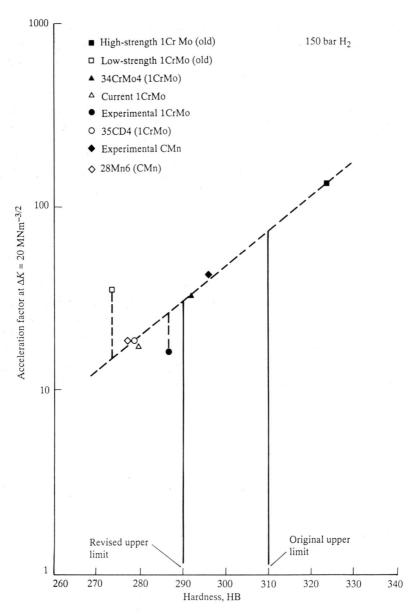

Figure 4.32   *The influence of hardness on the enhancement of fatigue crack growth by hydrogen (After Harris* et al.[21])

required to transmit high loads and provide long service lives under arduous fatigue conditions. Stemming mainly from the requirement of the aeroengine industry, major effort has been devoted to improving the level and consistency of bearing fatigue performance and, with the adoption of cleaner steelmaking techniques, it is claimed that bearing life has increased by a factor of 100 since the early 1940s.[22]

The grade of steel adopted internationally for through-hardened bearings is SAE 52100, the 1.0% C 1.5% Cr composition. This material is generally solution treated at a temperature of about 850°C, followed by oil quenching and tempering in the range 180–250°C. This results in a microstructure of lightly tempered martensite, primary (undissolved) carbides and up to about 5% retained austenite. For larger bearings, carburizing grades such as SAE 4720 (2% Ni–Mo) are adopted. In the United States, M50 (0.8% C, 4.0% Cr, 4.25% Mo) is used extensively in main shaft gas turbine bearings, helicopter transmission bearings and other aerospace applications. In 1983, Bamberger[23] introduced a variant of M50 steel which was designated M-50 NiL. This is a 0.12% C 3.5% Ni 4.0% Cr 4.25% Mo 1.2% V composition which is case hardened to produce bearings with high fracture toughness and long life.

Bearing fatigue testing

Whereas bearings are required to have long endurance lives, perhaps extending over several years, laboratory evaluation tests must be accelerated such that a meaningful result can be obtained in a matter of days. As such, these tests are undertaken at higher loads and speeds than those experienced under typical service conditions. Various bearing fatigue tests have been developed which are capable of assessing the performance of bearing steels in the form of balls, washers, cones and cylinders but in the UK the Unisteel washer test[24] is still in operation following its introduction in the early 1950s. A sectional view of the machine is shown in Figure 4.33. The test washer measures 76 mm o.d. × 51 mm

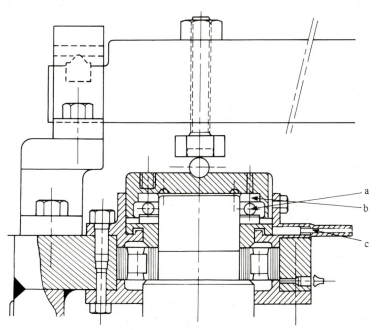

Figure 4.33 *Sectional view of Unisteel Bearing Fatigue Rig (a) balls; (b) test bearing; (c) standard thrust race (After Johnson and Sewell[24])*

i.d. × 5.5 mm thick and forms the top race of a standard thrust bearing. The cage of the standard bearing is operated with only nine balls instead of the normal 18 in order to provide a maximum calculated Hertzian stress of 3725 N/mm^2 at a relatively low load. Both sides of the specimen are tested.

Factors affecting fatigue performance

Johnson and Sewell[24] were among the earliest investigators to establish a good quantitative relationship between the bearing fatigue performance of SAE 52100 and inclusion content. The inclusions act as stress raisers, forming incipient cracks which then propagate under stress reversals until a fatigue pit (*or spall*) is formed on the surface of the component. As illustrated in Figure 4.34, Johnson and Sewell showed that oxide inclusions such as alumina and silicates have an adverse effect on fatigue performance, whereas sulphides appear to be beneficial. These authors showed that a better correlation between fatigue performance and inclusion content was obtained when titanium nitride was included in the inclusion count such that the inclusion parameter was based on:

$$\text{Number of alumina} + \text{silicate} + \frac{\text{TiN}}{2}$$

Oxides are considered to be detrimental because they are brittle and, as illustrated by Brooksbank and Andrews,[25] they also become surrounded by tensile stresses on cooling from elevated temperatures due to differences in the thermal expansion characteristics between the oxide particles and the matrix. The beneficial effect of sulphides is generally ascribed to the fact that they tend to encapsulate the more angular oxide inclusions, thereby reducing the detri-

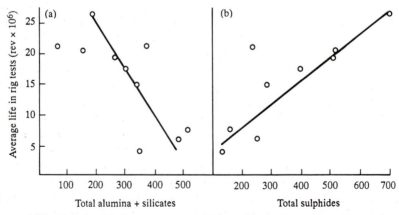

Figure 4.34 *Relationship between average life and inclusion content: counts based on total inclusions observed (× 750) in 516 fields representing a total area of ≃ 9 mm^2 (After Johnson and Sewell[24])*

mental tensile stresses. However, the beneficial effects of sulphides are often disputed and some hold the view that they are non-detrimental rather than positively beneficial. Other workers have indicated that TiN inclusions have a relatively small effect on bearing fatigue performance and such an effect is consistent with the tesselated stress calculations of Brooksbank and Andrews.

Given the importance of inclusion content in relation to bearing fatigue performance, major attention has been given to the development of reliable methods of inclusion assessment. For many years, both the steelmakers and bearing manufacturers have employed the Jerkontoret (JK) method which is specified in ASTM Practice for Determining the Inclusion Content of Steel – E45. This rates a steel in terms of the worst field of the 100 fields that are examined against standard charts. However, as stated by Hampshire and King,[26] this technique is capable of ignoring the fact that a steel with 100 equally bad fields would be ranked the same as a steel with only one bad field. Therefore a new method was required that took into account both inclusion severity and frequency and such a procedure is the SAM counting technique. This is detailed in Supplementary Requirements S2 of ASTM A 295-84. The SAM procedure concentrates on the frequency of *Type B* aluminates and *Type D* globular oxides because these tend to be the most damaging inclusions to bearing fatigue performance. *Type A* manganese sulphides are disregarded as being insignificant to fatigue life and *Type C* silicates are ignored because they occur infrequently in bearing steels.

The SAM count is regarded as a major improvement over the JK method but major strides have been made in introducing cleaner steels and a large number of fields must now be examined in order to obtain a statistically valid assessment of inclusion content. Attention is being given to the use of automatic image analysis techniques but these require very careful polishing procedures and, again, problems are being experienced with these techniques in the accurate assessment of steels with low oxygen and sulphur contents.

Relatively little work has been carried out on the role of primary carbides in bearing steels. In general, it is accepted that these particles are very beneficial in increasing the wear resistance of the material and they are also instrumental in inhibiting grain growth during solution treatment. However, it is considered that large primary carbides, particularly when present in segregated bands, act as stress raisers and are detrimental to fatigue life.

The high carbon content of through-hardening steels such as SAE 52100 depresses the M_f temperature of the material below room temperature, resulting in the formation of retained austenite. A similar situation also exists in the case of carburized bearings, the high alloy content augmenting the effect of carbon in depressing the M_s–M_f transformation range. However, the retained austenite will transform to martensite under the high Hertzian stresses induced in service and will tend to introduce dimensional changes in the bearing component. For this reason, a low level of retained austenite is desirable for critical bearing applications and is generally limited to a maximum of about 5%. The formation of strain-induced martensite will create a compressive residual stress and should therefore be beneficial to the fatigue performance. The *white bands* observed in the microstructure of SAE 52100 bearings after service are thought to be associated with the residual stresses created by the formation of strain-induced martensite.

Modern steelmaking methods

During the 1950s, vacuum degassing facilities were introduced which resulted in a marked improvement in the cleanness of bearing steels. By the early 1960s, argon shrouding of the molten stream was adopted which prevented reoxidation, leading to a further reduction in the non-metallic inclusion content. In the mid-1970s, many steelmakers introduced vacuum induction melting and vacuum arc remelting (VIM, VAR) to produce exceptionally clean steels. However, dramatic improvements in cleanness have also been obtained in bulk steelmaking processes for bearing grades, particularly with the introduction of secondary steelmaking facilities. Davies *et al.*[27] have described the facilities that have been installed at Stocksbridge Engineering Steels for the production of bearing steels and these are summarized in Table 4.12. Steels from a 100 t electric arc furnace are processed via a vacuum arc degassing (VAD) unit while melts from a 150 t furnace are transferred to a ladle furnace (LF), equipped with argon stirring, arc heating and alloying facilities. Care is also taken to minimize the inclusion content with the use of high alumina and graphitized magnesia refractories.

The SAM count rating for alumina (Type B) and globular oxides (Type D) on casts of SAE 52100, produced by Stocksbridge Engineering Steels between 1984 and 1986, have been monitored by Hampshire and King.[26] These data are

Table 4.12 *Facilities for the production of clean steels at Stocksbridge Engineering Steels*

Process route element	B Furnace route	C Furnace route
Slag-free tapping	Submerged taphole	Submerged taphole
Ladle stirring	Argon through sidewall plug	Argon through sidewall and bottom plugs
Slag control	Specific synthetic slags	Specific synthetic slags
Reheating	Arc heating in VAD	Arc heating in LF
Degassing	Integral part of VAD	Separate tank degasser
Alloying	via 13 alloy hoppers	via 13 alloy hoppers
Ladle refractories	75–85% Al_2O_3 walls and bottom 92% MgO slag line	82–85% Al_2O_3 walls and bottom 96% MgO slag line
Ingot teeming shroud	Inert gas	Inert gas
Ingot holloware	60% Al_2O_3	60% Al_2O_3
CC ladle to tundish[a]	Graphitized Al_2O_3 shroud	Not applicable
CC tundish	700-mm deep, MgO-lined	Not applicable
CC tundish to mould	Graphitized Al_2O_3 submerged entry shroud and flux, or inert gas	Not applicable

[a]CC = continuous caster.
After Davies *et al.*[27]

shown in Figure 4.35. The major decrease in Type D inclusions in April 1985 coincides with the introduction of the ladle steelmaking facility and it is reported that individual Type D SAM counts since that time have rarely exceeded a value of 3. On the other hand, the introduction of ladle steelmaking did not have a dramatic effect on the incidence of Type B alumina inclusions, the SAM count being between 3 and 8 over the period shown. Mean oxygen contents of less than about 10 ppm can now be obtained via the ladle refining route, i.e. approaching that produced by vacuum arc remelting.

High-speed steels

High-speed steels are so called because of their ability to maintain a high level of hardness during high-speed machining operations. They are characterized by high carbon contents, sometimes up to 1.5%, and major additions of strong carbide-forming elements such as chromium, molybdenum, tungsten and vanadium. Up to 12% Co is also included in some of the more complex grades. The constitution, manufacture and properties of high-speed steels have been reviewed very thoroughly in a book by Hoyle.[28]

Role of alloying elements

The microstructure of high-speed steels consists of primary and secondary carbides in a matrix of tempered martensite. The primary carbides are coarse particles which are not dissolved during solution treatment for the hardening operation and the secondary carbides are fine particles that are precipitated in martensite during tempering treatments. The role of the major alloying elements is outlined below.

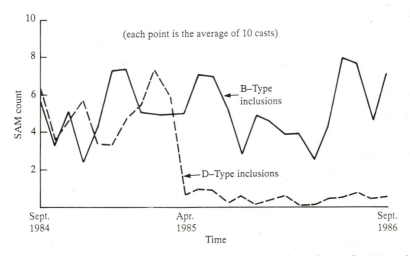

Figure 4.35 *SAM counts of steel delivered from major steel supplier (After Hampshire and King[26])*

A high carbon content is required in order to produce a hard martensitic matrix and also to form primary carbides. Both constituents are effective in providing abrasion resistance during metal cutting. However, the amount of carbon that can be accommodated in high-speed steels is limited on two counts. Firstly, carbon lowers the solidus temperature of steels substantially and therefore reduces the maximum temperature that can be employed for solution treatment prior to hardening. Secondly, carbon has a powerful effect in depressing the M_s–M_f temperature range and a very high level of carbon in solution will lead to excessive amounts of retained austenite.

The chromium content of most grades of high-speed steel is about 4%. Chromium forms carbides such as $M_{23}C_6$ and M_7C_3 but these carbides dissolve fairly readily and are taken into solution at the normal solution treatment temperatures employed for these steels, i.e. 1200–1300°C. Chromium is therefore added as a hardenability agent and promotes the formation of martensite. In addition, chromium is also beneficial in improving the scaling resistance of these steels at the high temperatures generated during machining operations.

Molybdenum and tungsten are the principal alloying elements in high-speed steels and can be present in levels up to about 10% and 20% respectively. From a metallurgical standpoint, the two elements are very similar and in most types of steel they are interchangeable on the basis of atomic weight %. Both elements are strong carbide formers and the principal carbide type is M_6C. This carbide is sometimes called eta-carbide and has a low solid solubility in steel. Therefore molybdenum and tungsten contribute little to the hardenability of high-speed steels but the small amounts that are dissolved are very effective in promoting tempering resistance and maintaining a high level of hardness at the cutting temperature (*red hardness*). Both elements also induce secondary hardening reactions in martensite, as illustrated schematically in Figure 4.36. This figure also indicates that appreciable softening does not take place until the temperature exceeds about 550°C, which therefore represents the effective maximum operating temperature for these steels.

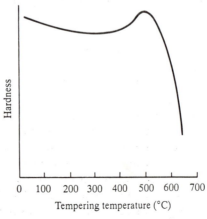

Figure 4.36   *Tempering curve for high-speed steel (schematic (After Hoyle[28])*

In the early development of high-speed steels, tungsten was regarded very much as the preferred element and molybdenum the substitute element to be tolerated only in special circumstances. However, this prejudice has now largely disappeared and there appears to be no technical reason for recommending a tungsten grade in favour of a comparable molybdenum grade.

All high-speed steels contain between 1% and 5% V and as the vanadium content is increased, the carbon is generally increased by at least 0.1% for each additional 1% V. Vanadium is a very strong carbide-forming element and produces extremely hard particles of V_4C_3 (MC type). This carbide has a very low solid solubility in steel and again contributes very little to hardenability. However, vanadium carbide improves the abrasion resistance of high-speed steels and is also beneficial as a grain-refining agent.

Although not a standard addition, up to 12% Co is incorporated in some high-speed steels and, according to Hoyle,[28] the role of this element is less clearly defined than that of the other major additions. Cobalt does not form carbides and has a high solubility in austenite. For a given hardening temperature, it reduces the level of retained austenite and also improves the secondary hardening response. Cobalt also increases the thermal conductivity, particularly at high temperatures. The net result is that cobalt increases the red hardness of high-speed steels and improves their performance in fast-cutting operations.

Heat treatment

Given the high carbon content and complex alloy design, it might be anticipated that the heat treatment of high-speed steels would involve rather complicated procedures. However, these procedures are based on sound metallurgical principles and are designed to achieve specific microstructural features and properties. A schematic illustration of the heat treatment of high-speed steels, including the annealing, hardening and tempering cycles, is shown in Figure 4.37.

Because of their high alloy content, high-speed steels are air hardening and will form martensite on cooling from the austenite temperature range. Therefore, after forging or hot-forming operations, these steels must be annealed in order to produce:

1. A softened condition for easier machining operations.
2. The relief of internal stresses.
3. A suitable microstructure for the subsequent hardening treatment.

Annealing is carried out by heating slowly to a temperature just above Ac_1 which involves temperatures in the range 850–900°C, depending upon the particular grade of steel. The material is held at the annealing temperature for two to four hours and then furnace cooled to a temperature below 600°C. The rate of cooling can then be increased. This treatment results in the formation of a ferrite matrix with finely dispersed carbide particles. However, the large primary carbides remain virtually unaffected by this treatment. After annealing, the hardness of the standard grades of high-speed steel is less than 300 HB.

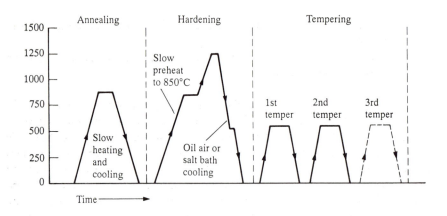

Figure 4.37   *The heat treatment of high-speed steels*

As indicated earlier, elements such as molybdenum, tungsten and vanadium form stable carbides which have only limited solubilities in steel. However, in order to produce a high-carbon martensite, with good tempering resistance and the facility for secondary hardening, it is essential that a proportion of these carbides is taken into solution. Solution treatment temperatures very close to the solidus temperatures are therefore employed, i.e. temperatures in the range 1200–1300°C, depending upon the grade. According to Hoyle,[28] the solution of $M_{23}C_6$ (Cr-based) carbide begins at temperatures just above 900°C and is complete at about 1100°C. The solution of M_6C (Mo- and W-based) carbides begins at about 1150°C and continues until the solidus is reached. On the other hand, MC (V-based) carbide is extremely stable and little solution is achieved, even at temperatures close to the solidus.

In order to minimize thermal shock, the steel is preheated slowly to a temperature of about 850°C. This is generally carried out in one furnace and the steel is then transferred to a high-temperature furnace with a neutral atmosphere. However, the soaking time at the hardening temperature must be short, e.g. two to five minutes, in order to minimize decarburization and grain growth.

Following solution treatment, the cooling operation can be carried out in air or by quenching into oil or a salt bath. Salt bath temperatures of 500–600°C are employed and the treatment is carried out in order to reduce temperature gradients and thereby reduce distortion and the risk of cracking. However, the salt bath treatment should only be long enough to allow the material to attain a uniform temperature and the component should then be air cooled. Transformation to martensite will begin at a temperature below 200°C and will be about 80% complete at room temperature. Whereas it is important to ensure that the steel reaches ambient temperature in order to achieve a high degree of transformation, tempering treatments must take place immediately after the hardening cycle in order to prevent stabilizing effects in the residual retained austenite.

Tempering treatments are carried out at 530–570°C and serve two purposes:

1. To produce a tempered martensitic structure which will be hard but stable at elevated temperatures.
2. To destabilize the retained austenite such that martensite will form on cooling to room temperature.

As indicated in Figure 4.36, a secondary hardening reaction takes place with a peak at a temperature of about 550°C, i.e. the typical tempering temperature for high-speed steel. Therefore, as well as providing a structure which will be stable in the short term up to this temperature, the tempering treatment also produces the maximum level of hardness. During the first tempering treatment, carbide precipitation takes place in both the martensite and retained austenite that were present in the structure after the hardening treatment. In the latter case, carbide precipitation depletes the phase in carbon and alloying elements, such that the M_s–M_f temperature range is raised and the retained austenite transforms to martensite on cooling to room temperature. Therefore at the end of the first tempering treatment, the microstructure will consist of:

1. Primary (undissolved) carbides;
2. Tempered martensite containing secondary carbides;
3. Untempered martensite;

and a second tempering treatment is required in order to temper the newly formed martensite. The second tempering treatment is again carried out at 530–570°C and for most grades this achieves the required microstructure, i.e. primary carbides in a matrix of tempered martensite. However, for optimum performance and maximum dimensional stability, it may be necessary to carry out a third tempering treatment, particularly in high-cobalt grades.

Standard specifications and uses

In the UK, high-speed steels are specified in BS 4659: 1989 *Tool and die steels*, which also includes details of hot-work tool steels, cold-work tool steels and plastic moulding grades. The specified ranges for the main alloying elements in high-speed steels, together with the hardness requirements in the annealed and hardened conditions, are summarized in Table 4.13. The grade designations adopted for the various types of tool steel in this specification follow the system laid down by the American Iron and Steel Institute (AISI) except that in all cases they are preceded by the letter 'B'. Thus grade T1 in the American standard is designated BT1 in the UK standard. In both cases, the letter 'T' refers to grades in which tungsten is the principal alloying element and the letter 'M' is used for molybdenum or tungsten–molybdenum grades. In the German system, the designations have the prefix 'S' (Schnellstahl) and the numbers that follow represent the levels of tungsten, molybdenum, vanadium and cobalt in that order. The chromium content is not specified but is most likely to be of the order of 4%. Thus M2/BM2 (6% W, 5% Mo, 4% Cr, 2% V) in the American and British standards have the same alloy design as S6-5-2 in the German standard.

Table 4.13 *High-speed steels – BS 4659: 1989. Details of composition and hardness*

Designation	C%	Cr%	Mo%	W%	V%	Co%	Hardness after annealing HB (max.)	Hardness after heat treatment HV (min.)
BM1	0.75–0.85	3.75–4.5	8–9	1–2	1–1.25	1 max.	241	823
BM2	0.82–0.92	3.75–4.5	4.75–5.5	6–6.75	1.75–2.05	1 max.	248	836
BM4	1.25–1.4	3.75–4.5	4.25–5	5.75–6.5	3.75–4.25	1 max.	255	849
BM15	1.45–1.6	4.5–5	2.75–3.25	6.25–7	4.75–5.25	4.5–5.5	277	869
BM35	0.85–0.95	3.75–4.5	4.75–5.25	6–6.75	1.75–2.15	4.6–5.2	269	869
BM42	1–1.1	3.5–4.25	9–10	1–2	1–1.3	7.5–8.5	269	897
BT1	0.7–0.8	3.75–4.5	0.7 max.	17.5–18.5	1–1.25	1 max.	255	823
BT4	0.7–0.8	3.75–4.5	1 max.	17.5–18.5	1–1.25	4.5–5.5	277	849
BT5	0.75–0.85	3.75–4.5	1 max.	18.5–19.5	1.75–2.05	9–10	290	869
BT6	0.75–0.85	3.75–4.5	1 max.	20–21	1.25–1.75	11.25–12.25	302	869
BT15	1.4–1.6	4.25–5	1 max.	12–13	4.75–5.25	4.5–5.5	290	890
BT21	0.6–0.7	3.5–4.25	0.7 max.	13.5–14.5	0.4–0.6	1 max.	255	798
BT42	1.25–1.4	3.75–4.5	2.75–3.5	8.5–9.5	2.75–3.25	9–10	277	912

After **BS 4659**: 1989.

Although 13 grades of high-speed steels are listed in BS 4659: 1989, the three popular grades of high-speed steel are listed in Table 4.14.

Table 4.14

Grade	C%	W%	Mo%	Cr%	V%
T1	0.75	18		4	1
M1	0.8	2	8	4	1
M2	0.85	6	5	4	2

In the United States, M10 (8% Mo, 4% Cr, 2% V) is also a common grade and with the three steels listed above makes up nearly 90% of the general purpose high-speed tools. However, an M10-type steel is not listed in the British standard.

Hoyle[28] has prepared a basic guide to tool selection and this is shown in Table 4.15. This selection reflects the benefit of cobalt additions in maintaining a high level of red hardness at high cutting speeds and the use of higher carbon and vanadium contents in order to increase the resistance to abrasive wear when cutting very hard materials.

Although high-speed steels constitute the major type of cutting tool, sintered carbides now feature very prominently in this market and ceramic materials, such as alumina and cubic boron nitride, are also being used for fast machining operations or for very hard workpiece materials.

Maraging steels

The term *maraging* relates to aging (precipitation strengthening) reactions in very low carbon martensitic steels for the development of ultra-high strengths, i.e. 0.2% proof strength values of 1400–2400 N/mm². Maraging steels are characterized by high nickel contents and a very important feature is that they exhibit substantially higher levels of toughness than conventional high-carbon martensitic grades of equivalent strength.

Table 4.15 *Selection of high-speed steels*

Application	Grade	Composition (%)					
Normal duty	M1	0.8C,	2W,	8Mo,	4Cr,	1V	
	M2	0.85C,	6W,	5Mo,	4Cr,	2V	
	T1	0.75C,	18W,	4Cr,	1V		
Higher speed cutting	M35	0.9C,	6W,	5Mo,	4Cr,	2V,	5Co
	T4	0.75C,	18W	4Cr,	1V,	5Co	
	T6	0.8C,	20W,	4Cr,	1.5V,	12Co	
Higher hardness	M15	1.5C,	7W,	3Mo,	5Cr,	5V,	5Co
	M42	1.3C,	9W,	3Mo,	4Cr,	3V,	10Co

Work on these steels began in the United States in the 1950s at the Bayonne Research Laboratory of the International Nickel Company and was directed primarily to the development of a high-strength material for submarine hulls. However, maraging steels proved unsuitable for this application and their main usage has been in the areas of aerospace, tooling and machinery, and structural engineering. An excellent summary of information on these steels is contained in a publication by the American Society for Metals.[29]

Metallurgy

Maraging steels generally contain about 18% Ni and the carbon contents are limited to 0.03% max. Typically, they are solution treated at a temperature of 820°C and air cooled to room temperature. This results in the formation of a martensitic structure, even in large section sizes due to the high hardenability effect conferred by the high nickel content and other alloying elements. In the solution-treated condition, the low-carbon martensitic structures provide the following range of properties:

- 0.2% proof strength, 790–830 N/mm^2
- tensile strength, 1000–1150 N/mm^2
- elongation, 17–19%
- reduction of area, 27–35%

These steels also contain substantial amounts of cobalt and molybdenum and smaller additions of titanium which promote age-hardening reactions when the solution-treated steels are aged at temperatures of about 480°C. In the 18% Ni–Co–Mo–Ti grades, the main precipitation-strengthening phases are Ni_3Mo (orthorhombic) and a complex sigma phase (tetragonal), based on FeTi.[30] As illustrated in Figure 4.38, the presence of cobalt intensifies the hardening effect of molybdenum but no aging effects are produced in high-cobalt–molybdenum-free compositions.[31]

Commercial grades

Although a large number of maraging grades has been developed, including compositions containing 20% and 25% Ni, the main commercial grades are based on the five composition ranges shown in Table 4.16. In this table, the numbers ascribed to the various grades relate to the nominal proof strength values in units of N/mm^2. However, these grades are also designated according to their proof strength values in ksi units, e.g. 18Ni 1700 (N/mm^2) = 18Ni 250 (ksi). As indicated in Table 4.16, the strength of the wrought grades is increased by increasing the levels of cobalt, molybdenum and titanium. Whereas the 18Ni 1400, 1700 and 1900 grades are aged for three hours at 480°C, the 18Ni 2400 grade requires three hours at 510°C or longer times at 480°C. The larger amount of titanium in 18Ni 2400 leads to a greater volume fraction of FeTi sigma phase precipitates compared with the lower strength grades.

Whereas the carbon content of these grades is restricted to low levels, the

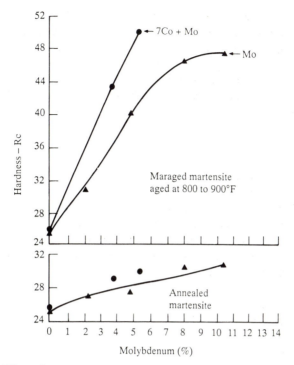

Figure 4.38 *Effect of molybdenum and cobalt on hardening response in maraging steels (After Decker* et al.[31]*)*

metallurgically active content is virtually zero due to the fixation of carbon by titanium. Elements such as silicon and manganese are also tightly controlled in order to promote high toughness levels. Aluminium is added primarily as a deoxidant, and although larger amounts will supplement the hardening reactions, this leads to a loss in toughness. Boron and zirconium are added in order to retard grain boundary precipitation, thereby improving toughness and stress corrosion resistance.[31]

Maraging steels are generally produced by vacuum melting and sometimes double vacuum melting (induction plus vacuum arc remelting) is employed. These procedures are designed to achieve the following:

1. Close control over the main alloying elements and the volatilization of impurities such as lead and bismuth.
2. Minimization of segregation.
3. Low levels of oxygen, nitrogen and hydrogen.
4. Low levels of inclusions.

As indicated in Figure 4.39, the toughness of the maraging grades is significantly better than that of tempered high-carbon martensitic steels.[33]

The casting grade 17Ni 1600 is normally solution treated by homogenizing for four hours at 1150°C, followed by air cooling. Maraging then takes place at

Table 4.16 Composition ranges for 18% Ni–Co–Mo maraging steels[32]

Grade	Wrought				Cast
	18Ni1400	18Ni1700	18Ni1900	18Ni2400	17Ni1600
Nominal 0.2% proof stress:					
N/mm² (MPa)	1400	1700	1900	2400	1600
tonf/in²	90	110	125	155	105
10³lbf/in²	200	250	280	350	230
kgf/mm²	140	175	195	245	165
hbar	140	170	190	240	160
Ni	17–19	17–19	18–19	17–18	16–17.5
Co	8–9	7–8.5	8–9.5	12–13	9.5–11
Mo	3–3.5	4.6–5.1	4.6–5.2	3.5–4	4.4–4.8
Ti	0.15–0.25	0.3–0.5	0.5–0.8	1.6–2	0.15–0.45
Al	0.05–0.15	0.05–0.15	0.05–0.15	0.1–0.2	0.02–0.1
C max.	0.03	0.03	0.03	0.01	0.03
Si max.	0.12	0.12	0.12	0.1	0.1
Mn max.	0.12	0.12	0.12	0.1	0.1
Si + Mn max.	0.2	0.2	0.2	0.2	0.2
S max.	0.01	0.01	0.01	0.005	0.01
P max.	0.01	0.01	0.01	0.005	0.01
Ca added	0.05	0.05	0.05	None	None
B added	0.003	0.003	0.003	None	None
Zr added	0.02	0.02	0.02	None	None
Fe	Balance	Balance	Balance	Balance	Balance

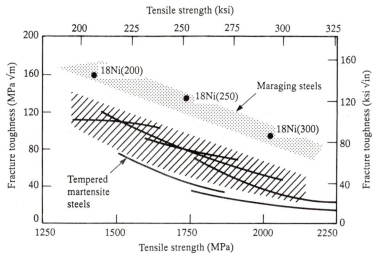

Figure 4.39 *Fracture toughness of maraging steels and other high-strength martensitic grades (After Magnee et al.[33])*

480°C for three hours, which is similar to the conditions used for the lower strength wrought grades. However, a more complex heat treatment is sometimes employed in order to improve toughness. This involves the following:

1150°C 4 hours + 595°C 4 hours + 820°C 1 hour + 480°C 3 hours

Corrosion behaviour

The corrosion rate of maraging steels in both marine and industrial atmospheres is almost half of that shown by high-strength, low-alloy steels. In sea water, both types of steels show similar corrosion rates initially, but after six months the maraging steels corrode more slowly. The International Nickel Co.[32] have published K_{ISCC} data for exposure in aqueous environments, with and without NaCl, which show that the 18% Ni maraging steels compare favourably with other high-strength steels.

Applications

From the foregoing remarks, it is apparent that maraging steels are capable of producing very high strengths with good fracture toughness but at a cost far in excess of that of conventional, low-alloy engineering steels. They are therefore used selectively in applications where weight saving is of paramount importance or where they can be shown to be more cost-effective than low-alloy grades. Some typical uses for maraging steels are listed in Table 4.17. As one might anticipate, these steels are used to advantage for weight reduction in aerospace and military applications but their excellent combination of properties, ease of

Table 4.17   *Applications for maraging steels[32]*

Aerospace	Tooling and machinery	Structural engineering and ordnance
Aircraft forgings (e.g. undercarriage parts, wing fittings)	Punches and die bolsters for cold forging	Lightweight portable military bridges
Solid-propellant missile cases	Extrusion press rams and mandrels	Ordnance components
		Fasteners
Jet-engine starter impellers	Aluminium die-casting and extrusion dies	
Aircraft arrestor hooks	Cold-reducing mandrels in tube production	
Torque transmission shafts	Zinc-base alloy die-casting dies	

Aircraft ejector release units	Machine components:
	    gears
	    index plates
	    lead screws	

heat treatment and dimensional stability also offer attractions in the more commercial sector of tooling and machinery.

Steels for steam power turbines

Electricity is generated on a large scale by the following sequence of operations:

1. The production of steam at high temperature and pressure in fossil-fired boilers or nuclear reactors.
2. The passage of the steam through a turbine where it impinges on the blades of a rotor, thereby creating rotational energy.
3. The transmission of the energy developed in the turbine rotor to a generator rotor which produces electricity in the windings of the stator.

The rotor shafts in the turbine and generator are produced from very large, high-integrity forgings which operate at speeds of about 3000 rev/min and over a range of temperatures, depending on their position in the power train. In a typical 660 MW coal-fired station, the turbine has three stages, namely high pressure (HP), intermediate pressure (IP) and low pressure (LP), and the last stage can involve two or three cylinders. On the other hand, the turbines for large water-cooled reactors operate at 1200 MW and involve only the HP and LP stages with very large rotors. This is illustrated in the data in Table 4.18 by Collier and Gemmill[34] for the weights of finished forgings. These authors also point out that the ingot weights for these components are usually a factor of two or three times that of the finished forgings and often involve the combined production of several steelmaking and secondary refining units in order to achieve the required ingot weights.

Table 4.18

Component	Weight (tons)	
	660 MW Fossil unit	*1200 MW Nuclear unit*
HP rotor	15	90
IP rotor	30	–
LP rotor	50	210
Generator rotor	90	190

Attention will be focused on the rotors for turbines and generators but reference will also be made to other components such as casings, bolts and blades. A very detailed text on the materials used in both steam and nuclear power plant has been prepared by Wyatt.[35]

Turbine casings

The casings for the HP and IP stages of turbines are made from large castings, and in the UK the 0.5% Cr–Mo–V grade is favoured:

0.15% C, 0.5% Cr, 0.5% Mo, 0.25% V

The casings are generally cast in two halves which are bolted together longitudinally along a heavy flange. In fossil-fired plant, steam inlet temperatures of 540 or 565°C are employed and therefore the casings are subjected to high internal pressure within the creep range. In addition, the *two shifting* operation of being on load during the day and shut down overnight imposes significant thermal fatigue stresses on the casing and other components in the HP and IP stages.

Repair welding in 0.5% Cr–Mo–V casings can give rise to stress relief cracking which is associated with low creep ductility in the coarse-grained heat-affected zone of weldments. This problem is generally overcome by adopting titanium as opposed to aluminium deoxidation and this practice has proved to be beneficial to both weldability and rupture ductility.

HP and IP rotors

The following 1% Cr–Mo–V composition is used world-wide for the rotor shafts of both the HP and IP stages of turbines:

0.25% C, 1.0% Cr, 1.0% Mo, 0.3% V

As indicated earlier, the HP rotor is the smallest in the turbine and operates at temperatures up to the main steam temperature of 540–565°C. During operation, the rotors are subjected to centrifugal forces and also to torques which mainly affect the rotor journals due to their smaller cross-sectional area. However, given the temperature of operation, the creep and rupture behaviour of the HP rotor is of major concern.

The IP rotor is similar in design but larger than the HP rotor and again there is the need for high creep and rupture strength in the hotter regions. However, at the exhaust end of the IP stage, the blades are longer and impose greater centrifugal forces. This factor, coupled with the cooler operating conditions, therefore introduces fracture toughness as an important design parameter. In addition, transient operating phases, such as start-up and shut-down, can result in further thermal stresses and reinforce the need for adequate fracture toughness at lower temperatures. However, there tends to be an inverse relationship between rupture strength and toughness and whereas the best creep and rupture strengths in 1% Cr–Mo–V steel are obtained in an upper bainitic microstructure, such material has poor toughness. On the other hand, the lower bainitic or martensitic structures that promote good fracture toughness have poor creep strength. The final hardening treatment for 1% Cr–Mo–V rotors therefore depends very much on whether high-temperature strength or good toughness is considered to be the more important design criterion. In the UK, rotors are oil quenched for improved fracture toughness, whereas in the United States, air cooling is adopted in favour of creep strength.

Both Reynolds *et al.*[36] and Viswanathan and Jaffe[37] have reviewed the factors that might lead to an improved combination of creep strength and fracture toughness. These include the use of vacuum carbon deoxidation, electroslag remelting and modifications to the traditional 1% Cr–Mo–V composition to incorporate higher levels of chromium (1.5%) and an addition of nickel (0.7%). It has also been shown that improvements in the toughness of HP and IP rotors can be obtained by reducing the level of impurity elements and suppressing temper embrittlement. However, it is generally accepted that temper embrittlement is more pronounced in the 3.5% Ni–Cr–Mo–V LP rotor grade and this matter will receive greater attention in the next section.

LP rotors

The LP rotor is the largest in the turbine assembly and operates at temperatures between 270°C and ambient. The blades in this segment are also very long, e.g. up to 1.12 m, and impose high stresses on the rotor shaft. Therefore the main design criteria in LP rotors are a high proof strength and good fracture toughness as opposed to the major requirement for high creep strength in the HP and IP components. The 0.2% proof strength values for LP rotors are of the order of 750 N/mm² and the increasing demand for good fracture toughness is illustrated by the data in Table 4.19 for FATT values over the period from 1970 to 1985.

Table 4.19

Period	FATT °C[a]	
	Average	*Range*
1970–1975	+25	0 to +50
1976–1980	+16	−18 to +40
1981–1985	+10	−4 to +20

[a]Fracture appearance transition temperature.

The material used for LP rotor forgings is 3.5% Ni–Cr–Mo–V steel to the following composition:

0.25% C, 3.5% Ni, 1.5% Cr, 0.5% Mo, 0.1% V

In addition, the steel is made to low levels of silicon (0.1% max.) and manganese (typically 0.2%), together with restricted levels of elements such as arsenic, antimony, tin and phosphorus in order to minimize the effects of temper embrittlement. The evolution of the current composition will be described briefly to reflect the various changes that have taken place to accommodate the demands for increased turbine size and improved fracture toughness.

Boyle *et al.*[39] have given a very detailed account of the development work that was undertaken in the United States in the 1950s following failures in a number of turbine and generator rotors. These failures were attributed to inadequate fracture toughness and prompted an industry-wide development to improve the performance of the 2.7% Ni 0.5% Mo 0.1% V steel that was used for large turbine rotors at that time. It became evident that temper embrittlement was a significant factor in contributing to low toughness and the manganese content of the steel was reduced from 0.7% to 0.4%. In order to compensate for the loss in hardenability and to improve toughness, the nickel content was raised from 2.7% to 3.5%. Boyle *et al.* state that nickel contents as high as 5% were investigated but that the benefits derived from additions greater than 3.5% were insufficient to compensate for the difficulties introduced by the depression of the Ac_1. Presumably these difficulties related to the fact that the Ac_1 can be depressed to a temperature close to or below the nominal tempering temperature with the potential risk of reaustenitization.

Although changes to the manganese and nickel contents realized significant improvements in toughness, there was also the need to develop a composition for larger diameter rotors which required higher tensile and yield strengths, together with good toughness and ductility. This led to the addition of 1.5% Cr to the modified Ni–Mo–V to provide the current 3.5% Ni–Cr–Mo–V grade. Since the 1950s, a considerable amount of work has been carried out in the UK to improve the fracture toughness of LP and generator rotors, particularly in relation to the suppression of temper embrittlement. In addition, the original FATT data have been augmented with K_{IC} fracture toughness values and a good correlation between the two parameters has been derived via the expression:

$$K_{IC}MNm^{-3/2} = \frac{6600}{60 - B}$$

where $B = (\text{test temperature} - \text{FATT value})°C$.

Following homogenization treatments at temperatures around 1150°C, rotors are heated slowly to a solution treatment temperature of 840°C and then lowered vertically into an array of high-pressure water sprays. Modern LP rotors have a diameter in excess of 1700 mm and the above treatment results in the formation of a rim structure which is essentially martensitic and a core structure which is essentially bainitic. In order to avoid quench cracking, the large forgings are not quenched directly to room temperature but are held for a period at a temperature of 200°C in order to reduce internal stresses. The material is then tempered in the range 560–640°C, the lower bound avoiding the formation of isothermal temper embrittlement and the upper bound avoiding the risk of partial reaustenitization. In turn, the rate of cooling adopted from the tempering temperature represents a compromise between the fast cooling that inhibits temper embrittlement and the slow cooling that minimizes internal stress. The compromise of fan cooling is usually employed.

At one time, the requirement for very low levels of silicon and manganese would have posed major problems in deoxidation but these have been overcome by the introduction of vacuum carbon deoxidation (VCD). VCD treatment is also thought to be beneficial in improving the solidification characteristics and reducing the segregation effects in large forging ingots.

Turbine generator end rings

The copper windings in an electrical generator are not strong enough to resist the rotation stress and *end rings* are fitted to retain the windings. These components are very highly stressed and have 0.2% proof strength values greater than 1200 N/mm². A further requirement is that the material should be non-magnetic in order to minimize electrical losses from the generator. Up until the late 1970s, a cold-worked steel containing 18% Mn 5% Cr was used worldwide for this application but the main disadvantage of this material was its susceptibility to stress corrosion cracking (SCC). This could arise in moist conditions and the threat of catastrophic disintegration during service necessitated frequent inspection.

Because of these problems, attention was turned to the development of a more SCC-resistant material, leading to the introduction of a steel of the following composition:

0.08% C, 0.3% Si, 18.0% Mn, 18.0% Cr, 0.60% N

This material is substantially solid solution strengthened by nitrogen and, additionally, this element also serves to preserve the austenitic (and therefore non-magnetic) structure in the presence of a high chromium content. The 18% Mn 18% Cr steel also offers the facility for attaining higher 0.2% proof strength values than 18% Mn 5% Cr steel, i.e. values up to 1400 N/mm².

18% Mn 18% Cr steel now represents the first-choice material for end rings and after 10 years trouble-free service has virtually eliminated the need for in-service inspection

Turbine bolts

Large bolts or studs are used to maintain steam tightness in the flanges of high-temperature turbines. These are tightened to a tension that imposes an elastic strain of about 0.15% but at the high operating temperatures this tension reduces due to the process of creep (*stress relaxation*). Therefore bolting materials must have good creep strength in order to minimize relaxation and maintain steam tightness. Repeated relaxation and retightening will lead eventually to creep fracture and therefore the utilities limit both the operating period and the number of retightening operations so as to avoid fracture during service.

In the UK, the majority of the bolting requirements are satisfied by a range of Cr–Mo and Cr–Mo–V steels and the evolution of these steels has been described by Everson *et al.*[40] Details of the composition, heat treatment and high-temperature properties of these steels are given in Table 4.20. The Cr–Mo (Group 1) steel was introduced in the 1930s when the steam temperatures were of the order of 450°C. By the late 1940s, steam temperatures had risen to about 480°C and higher strength bolts were required. This led to the introduction of the Cr–Mo–V (Group 2) steel in which the higher strength is achieved by the formation of a fine precipitate of V_4C_3 on tempering at 700°C. However, steam temperatures continued to rise in the UK in pursuit of higher operating efficiency and in 1955 reached a level of 565°C. This change necessitated a further increase in strength and this led to the introduction of the 1% Cr 1% Mo 0.75% V (Group 5) steel. However, this material proved to have poor rupture ductility due to intergranular cracking after only short exposure at the operating temperature. A major research programme was therefore undertaken on this problem and this led to the development of the Cr–Mo–V–Ti–B (Group 6) steel. This composition represented the addition of about 0.1% Ti and 0.005% B to Group 5 steel which produced a significant improvement in rupture ductility. According to Everson, Orr and Dulieu, the principal effect is due to boron, about 50% of which is incorporated in the V_4C_3 precipitates and the remainder is dissolved in the matrix. This produces a stabilizing effect on the V_4C_3 near the grain boundary regions, making the carbides more resistant to dissolution and so reducing the rate at which denuded zones are formed. Titanium is added primarily as a nitrogen-fixing agent and so preventing the formation of boron nitride, which is metallurgically inactive. However, the formation of TiN leads to refinement of the austenite grains which also contributes to improved rupture ductility. More recently, the Cr–Mo–V–Ti–B steel has been used as boiler support rods, operating at temperatures up to 580°C.

Nimonic 80A (Ni 20.0% Cr, 2.4% Ti, 1.4% Al) is significantly stronger than the low-alloy steels and allows the use of smaller bolts and more compact

Table 4.20 *UK turbine bolting steels*

SES[a] Trade name	CEGB Code	BS 1506:1986 Type no.	Nominal composition (%)					Typical heat treatment	RT tensile strength (N/mm^2)	Strength			
			C	Cr	Mo	V	Others			Rupt.		Relax[b]	
										c	d	c	d
DUREHETE 900	GP 1 (Cr–Mo)	631–850	0.4	1	0.5	–	–	870°C OQ+660°C	850–1000	234	97	81	–
DUREHETE 950	GP 2 (Cr–Mo–V)	671–850	0.4	1	0.5	0.25	–	950°C OQ+700°C	850–1000	324	151	83	–
–	GP 3 (3Cr–Mo–V)	–	0.3	3	0.5	0.75	–	–	–	–	–	–	–
–	GP 4 (Mo–V)	–	0.2	–	0.5	0.25	–	–	–	–	–	–	–
DUREHETE 1050	GP 5 (1Cr–Mo–V)	–	0.2	1	1	0.75	–	Superseded by Durehete 1055					
DUREHETE 1050	GP 6 (1Cr–Mo–V)	681–820	0.2	1	1	0.75	Ti, **B**	980°C OQ+700°C	820–1000	418	280	141	70

[a] SES, Stocksbridge Engineering Steels.
[b] Stress relaxation for 0.15% strain.
[c] 10^4 h values (N/mm^2) at 500°C.
[d] 10^4 h values (N/mm^2) at 550°C.

After Everson et al.[40]

flanges than are possible with ferritic bolting materials. It is also likely that Nimonic 80A would replace Cr–Mo–V (Group 6) steel as the principal turbine bolting material if steam temperatures were to be raised above the present operating level of 565°C.

Detailed studies have also shown that marked improvements in the rupture characteristics of Cr–Mo–V bolting steels can be produced by restricting the level of residual elements. Thus the rupture life of Cr–Mo–V steel can be related to the 'R' factor (P + 2.43As + 3.75Sn + 0.13Cu) and Cr–Mo–V bolting steels are now made to a typical 'R' value of 0.07% compared to 0.2% in early years. This has led to a substantial increase in rupture ductility. The 'R' value can also be reduced to a level of 0.015 in VIM–VAR material to provide further improvements in creep strength and ductility.

Turbine blades

Apart from locations at the extreme ends of steam turbines, all the blades are made in 12% Cr steels. Although the metallurgy of these steels will be discussed in the next chapter, for the sake of continuity the use of 12% Cr steels in turbine blades is better described at this stage.

In the HP turbine, the blades are short and operate under the maximum steam temperature, i.e. 565°C. The creep strength of 12% Cr steels is not adequate to operate at such temperatures and therefore the first few rows of blades at the HP inlet are generally manufactured from Nimonic 80A. In the LP turbine, the blades are long, and in large turbines the exhaust blades can exceed a length of one metre. Such blades generate high centrifugal forces and again 12% Cr steels are not strong enough to cope with the conditions imposed. In such situations, the precipitation-strengthened FV520B steel may be employed, which has the composition shown in Table 4.21.

Table 4.21

C %	Ni %	Cr %	Mo %	Cu %	Nb %
0.05	5.5	14	1.6	1.5	0.3

However, in recent years, interest has grown in the use of titanium alloys, typically Ti 6% Al 4% V (Ti6Al4V), as a substitute for martensitic stainless steels in the outlet stages of turbines. The reduced weight and high resistance to corrosion fatigue of the titanium alloys allows the length of the blades to be increased substantially, compared to what can be achieved in 12% Cr steels. Thus the exhaust area of the LP turbine can be increased by about 50%, which increases the power by the same amount.

For all stages between the inlet and outlet, the grades of 12% Cr steels shown in Table 4.22 are employed. The tempering resistance and creep strength of

Table 4.22

Type	C %	Si %	Mn %	Ni %	Cr %	Mo %	V %	Nb %
12Cr–Mo	0.1	0.3	0.3	–	12.5	0.75	–	–
12Cr–Mo–V	0.1	0.3	0.6	0.8	12	0.6	0.2	–
12Cr–Mo–V–Nb	0.13	0.5	1	0.8	11.2	0.6	0.2	0.4

these materials is increased progressively with the addition of molybdenum, vanadium and niobium to the 12% Cr base. As illustrated in the next chapter, nickel is added to such steels in order to preserve an austenitic structure at high temperatures, in the presence of ferrite-forming elements such as molybdenum, vanadium and niobium.

Although the 12% Cr steels have adequate resistance to attack in moist air, the formation of water droplets in the final stages of the turbine can result in severe erosion problems. This is overcome by brazing strips of erosion-resistant materials such as stellite (cobalt-based) to the leading edges of LP blades.

Medium–high-carbon pearlitic steels

Although most of the steels that are used in engineering applications are heat treated to form a tempered bainitic/martensitic structure, there are notable examples in which the required strength is generated in air-cooled, medium–high-carbon steels with a predominantly pearlitic microstructure. These include micro-alloy forging grades, rail steels and high-carbon wire rod and these applications will be discussed in the sections that follow. However, as a precursor to these discussions, it is worthwhile to review very briefly the physical metallurgy of pearlitic steels.

In Chapter 3, the Hall–Petch relationship was introduced, namely:

$$\sigma_y = \sigma_i + k_y \, d^{-\frac{1}{2}}$$

where σ_y = yield strength
σ_i = friction stress which opposes dislocation movement (Peierls stress)
k_y = dislocation locking term
d = ferrite grain size (mm)

It was stated that this basic relationship was extended later to take account of the solid solution strengthening effects of alloying elements and expressed as follows:

$$\sigma_y = \sigma_i + k' \, (\% \text{ alloy}) + k_y \, d^{-\frac{1}{2}}$$

Following this concept, Pickering and Gladman[41] then produced the following quantitative relationships for yield and tensile strength:

$$\text{YS (N/mm}^2\text{)} = 53.9 + 32.3\% \text{ Mn} + 83.2\% \text{ Si} + 354 \text{ N}_f + 17.4d^{-\frac{1}{2}}$$

$$\text{TS (N/mm}^2\text{)} = 294 + 27.7\% \text{ Mn} + 83.2\% \text{ Si} + 3.85\% \text{ pearlite} + 7.7d^{-\frac{1}{2}}$$

These equations were derived by statistical analysis of steels containing up to 0.25% C and one of the interesting points to emerge was the absence of a term for carbon or pearlite in the equation for yield strength.

In the 1970s, Gladman *et al.* went on to develop equations for steels containing up to 0.9% C[42] and their equation for yield strength was presented in the following form:

$$\sigma_y = f_\alpha^n \sigma_\alpha + (1 - f_\alpha^n)\sigma_p$$

where σ_y = yield strength of the ferrite–pearlite aggregate
σ_α = yield strength of ferrite
σ_p = yield strength of pearlite
f_α = volume fraction of ferrite

and by implication $(1 - f_\alpha)$ = volume fraction of pearlite.

Thus the yield strength of the aggregate is presented as the sum of the separate contributions from ferrite and pearlite and weighted according to their volume fractions. In the absence of pearlite, $\sigma_y = \sigma_\alpha$ and in the absence of ferrite, $\sigma_y = \sigma_p$. The index n in the above equation allows the yield strength to vary with pearlite content in a non-linear manner and, as indicated below, it was given the value of $\frac{1}{3}$. The full quantitative equations for yield and tensile strengths are as follows:

$$\text{YS (N/mm}^2\text{)} = f_\alpha^{\frac{1}{3}} [35 + 58.5\% \text{ Mn} + 17.4d^{-\frac{1}{2}}] + (1 - f_\alpha^{\frac{1}{3}}) [178 + 3.85S_0^{-\frac{1}{2}}]$$
$$+ 63.1\% \text{ Si} + 426 \sqrt{\%\text{N}}$$
$$\text{TS (N/mm}^2\text{)} = f_\alpha^{\frac{1}{3}} [247 + 1146 \sqrt{\%\text{N}} + 18.2d^{-\frac{1}{2}}] + (1 - f_\alpha^{\frac{1}{3}}) [721 + 3.55S_0^{-\frac{1}{2}}]$$
$$+ 97.3\% \text{ Si}$$
where S_0 = interlamellar spacing of pearlite (mm)

Thus in high-carbon steels the volume fraction of pearlite $(1 - f_\alpha)$ and the interlamellar spacing (S_0) have a significant effect on both the yield and tensile strengths. The major difference between the two equations is the value of the constants, namely 35 and 178 N/mm² for yield strength and 247 and 721 N/mm² for tensile strength. Manganese in solid solution appears to have no significant effect on tensile strength, but this element appears in the equation for yield strength. The components of yield strength in medium- to high-carbon steels, containing 0.9% Mn, 0.3% Si and 0.007% N, are shown in Figure 4.40. This indicates that ferrite makes a significant contribution to the yield strength even when there is 90% pearlite in the microstructure.

Gladman *et al.*[42] also generated an equation for the impact transition temperature of high-carbon steels and this is shown below:

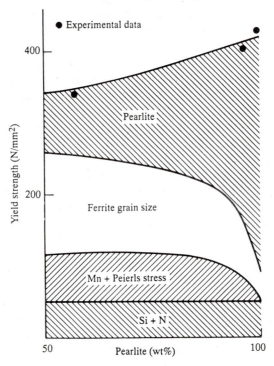

Figure 4.40   *Components of yield strength in high-carbon steels (After Gladman et al.[42])*

$$27 \text{ J ITT } (°C) = f_\alpha [-46 - 11.5d^{-\frac{1}{2}}]$$
$$+ (1 - f_\alpha)[-335 + 5.6S_o^{-\frac{1}{2}} - 13.3p^{-\frac{1}{2}} + 3.48 \times 10^6 t]$$
$$+ 48.7\% \text{ Si} + 762 \sqrt{\%N}$$

where d = ferrite grain (mm)
    p = pearlite colony size (mm)
    t = cementite plate thickness (mm)

This equation again emphasizes the importance of a fine grain size in producing a low impact transition temperature and it should be noted that the coefficient for the pearlite colony size p (13.3) is of a similar order to that of the ferrite grain size d (11.5).

Rail steels

Up until the 1970s, rails for passenger and freight trains were regarded as relatively simple undemanding products and the specifications had changed very little for a number of decades. However, investment in railway systems, the advent of high-speed passenger trains and the requirement for longer life track imposed a demand for rails of high quality, greater strength and tighter geometric tolerances. Therefore there have been major innovations in the past 20 years in terms of method of manufacture, degree of inspection and range of products.

Typically, rail steel is produced in large BOS vessels and is vacuum degassed prior to being continuously cast into large blooms. Vacuum degassing, coupled with ladle trimming facilities, permits very tight control over chemical composition. After casting, the blooms are placed in insulated boxes, whilst still at a temperature of about 600°C, and are cooled at a rate of 1°C per hour for a period of three to five days. This treatment, coupled with prior vacuum degassing, reduces the hydrogen level in the finished rail to about 0.5 ppm, thereby reducing substantially the susceptibility to hydrogen cracking. The blooms are then reheated and rolled directly to the finished rail profile. The rail produced from each bloom is hot sawn to specific lengths prior to passage through a rotary stamping machine *en route* to the cooling areas. Depending upon the properties required, the rails are either cooled normally in air or subjected to enhanced cooling for the development of high strength. On cooling to room temperature, the rails are passed through a roller-straightener machine which subjects the section to a number of severe bending reversals and emerge with a very high degree of straightness. Finally, the rails pass through a series of ultrasonic, eddy current and laser inspection stations which monitor non-metallic inclusions, external defects and the flatness of the running surface. The final operation is the cold sawing of the rail ends on high-speed machines with carbide-tipped blades. Rails are generally supplied in lengths up to 36 m.

Rail steel specifications

Historically, high-volume rail steels have been based on fully pearlitic microstructures which are characterized by high resistance to wear and plastic flow, both of which are major property requirements for good rail performance. Although the potential of martensitic structures has been evaluated, they proved to be unsuitable, possessing inadequate toughness and ductility. Therefore rail steels continue to be based on pearlitic microstructures which are generated through various combinations of carbon, manganese and other elements.

Rail steel specifications can be classified into three types, based on tensile strength:

1. Normal grades, ~ 700 N/mm^2 min. TS
2. Wear-resisting grades, 880 N/mm^2 min. TS
3. High-strength grades, 1080–1200 N/mm^2 TS

Normal grades

Typical examples of *normal grades* are BS 11: 1985 Normal and UIC 860-O Grade 70 (Table 4.23). These are the high-tonnage grades which are used in normal service conditions in conventional railways, including high-speed passenger traffic (200 km/h), and medium-speed (100 km/h) relatively heavy axle load (25 tonne) freight. The majority of London Transport underground track is also laid in BS 11 Normal grade.

Table 4.23

Grade	C%	Si%	Mn%	TS min. (N/mm²)	Elong. min. 5.65 $\sqrt{S_o}$
BS 11 Normal	0.45–0.6	0.05–0.35	0.95–1.25	710	9
UIC 860-0 Grade 70	0.4–0.6	0.05–0.35	0.8–1.25	680	14

Wear-resistant grades

The hardness and wear resistance of pearlitic steels are increased by refining the pearlite lamellae. This is achieved by increasing the carbon and manganese contents as illustrated in the specifications for *wear-resisting grades* given in Tables 4.24 and 4.25.

Table 4.24 *BS11 wear-resisting grade A/UIC 860-0 Grade 90A*

C%	Si%	Mn%	TS min. (N/mm²)	Elong. min. 5.65 $\sqrt{S_o}$
0.65–0.8	0.1–0.5	0.8–1.3	880	8

Table 4.25 *BS11 wear-resisting grade B/UIC 860-0 Grade 90B*

C%	Si%	Mn%	TS min. (N/mm²)	Elong. min. 5.65 $\sqrt{S_o}$
0.55–0.75	0.1–0.5	1.3–1.7	880	8

Thus similar levels of tensile strength can be obtained from various combinations of carbon and manganese, both of which depress the temperature of transformation from austenite to pearlite and thereby refine the pearlite lamellae. These wear-resisting grades are used for heavy axle loads, high-density traffic routes or tightly curved track. However, the use of wear-resisting rails on conventional railways can also show economic advantages.

High-strength grades

For the extremely arduous service conditions encountered in tightly curved track and under very high axle loads, even higher strengths are required which demand further refinement of the pearlitic structure. Up until 1985, most European railmakers produced such material by adding up to 1% Cr to the basic C–Mn composition and this increased the hardness of the rail head from 280 BHN to approximately 330 BHN with an improvement factor of about two in wear resistance. However, the gradual replacement of bolted track by welded track brought about the requirement for an adequate level of weldability, a property not readily satisfied with the high hardenability introduced by the addition of 1% Cr. Attention therefore turned to the use of accelerated cooling from the austenite range, rather than high-alloy additions, for the depression of the pearlite transformation and the development of high-strength rails. A number of railmakers have now introduced in-line cooling for the head hardening of rails and detailed accounts of the computerized facility installed by British Steel at Workington are given in publications by Preston[43] and Hodgson and Preston.[44] Brief details are as follows.

The rail leaves the finishing stands at a temperature of about 1000°C and is stood head up. The rail passes under temperature monitors and then into a 55-m-long cooling train where it is sprayed with water over all surfaces. The outgoing rail leaves the cooling station at dull red heat and the main transformation to pearlite takes place in still air. Obviously the rate of cooling has to be balanced very finely so as to provide sufficiently high rates of cooling in the centre of the rail head so as to depress the transformation of austenite to pearlite without cooling the surface of the rail at a rate which would lead to the formation of bainite or martensite. During cooling, the rail is driven through the cooling station by rollers which also maintain straightness in the rail. On leaving the cooling station, the rail is finally turned on its side and passes down the cooling banks for finishing in the normal manner. Using this facility, the hardness developed in the head of a 0.8% C 0.9% Mn rail is controlled to within a fairly narrow range of about 350–370 BHN.[44]

Flash butt welding is used to join lengths of rail into continuous track and Preston[43] has commented on the effects of this process on the hardness of enhanced cooled rail. In flash butt welding, the rail ends are heated to high temperatures by an electric arc and then squeezed together by a hydraulic force of 40–60 tonne. The rail ends fuse, excess material is extruded from the joint and material in the heat-affected zone cools through the transformation range at the rate of about 1.3°C/s. In contrast, the head of mill-hardened rail cools at a rate of about 2.5°C/s, and after flash butt welding, the hardness of enhanced cooled 0.8% C 0.9% Mn steel falls to about 300 BHN compared with 350–370 BHN in the mill-hardened condition. However, this can be avoided by increasing the alloy content of the steel such that a hardness of about 370 BHN is developed under a natural cooling rate of 1.3°C/s after welding.

Wear resistance of rails

British Steel Technical has carried out extensive laboratory wear tests on rail steels at its Swinden Laboratories in Rotherham. This involves the rotation of discs of railway wheel and rail materials, under a controlled contact stress and with a controlled amount of slip between the two discs. Wear is determined by weight loss on the rail test disc and expressed in terms of mg/m of slip. In this test, a very clear relationship has been established between wear rate and hardness and this relationship is shown for a range of rail steel grades in Figure 4.41.[45] Thus BS 11 with the lowest hardness and tensile strength exhibits the highest wear rate but this falls progressively to low values as the alloy content and hardness of the materials is increased. However, whilst producing the correct order of merit, this test exaggerates the relative performance of the various steels under operating conditions and therefore laboratory rig testing has to be supplemented by measurements of rail wear under normal track service. An illustration of the development of *side wear* in BS 11 Normal and 1% Cr rails is given in Figure 4.42. This indicates that 1% Cr material has a wear resistance of three to four times that of BS 11 Normal under the particular track conditions identified in this figure. Other track tests indicate that wear-resistant grades have twice the resistance of BS 11 Normal.

Side wear of the type illustrated in Figure 4.42 is a function of axle load and track curvature, the centrifugal force causing the flange of the wheel to scrub against the gauge corner head of the high outer rail in the curve.

Austenitic 14% Mn rails

When the service conditions are such that exceptionally high rates of wear are

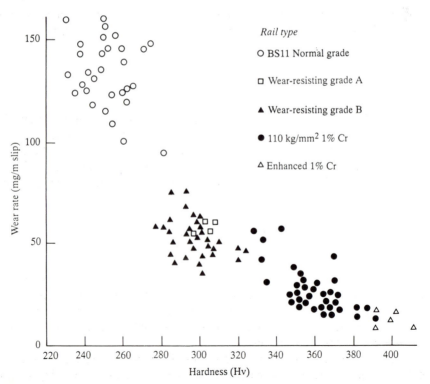

Figure 4.41   *Effect of hardness on wear rate in laboratory tests (After British Steel[45])*

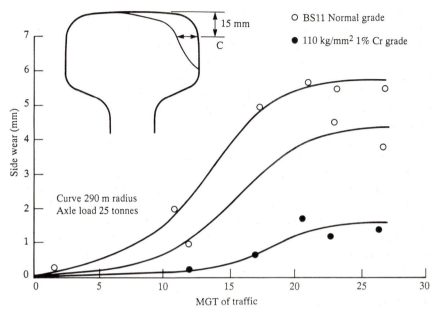

Figure 4.42 *Side wear in normal and 1% Cr rails under identical service conditions (After British Steel[45])*

experienced in high-strength rails, then consideration is given to the use of rails of the following composition which develop an austenitic microstructure:

C%	Si%	Mn%
0.75–0.9	0.2–0.4	13–14

The metallurgy of this material is complex but such steel has a very high resistance to wear because of its high rate of work hardening when subjected to applied stress or abrasion. This special grade of steel is made in electric arc furnaces but is rolled to rail in the same type of mill as that employed for the pearlitic grades. The mechanical properties[45] of 14% Mn rail material are given below:

TS (N/mm²)	YS (N/mm²)	Elong. (%)
818–973	355–386	40–60

The low YS/TS ratio illustrated above is indicative of the high rate of work hardening and material with an initial hardness of 180–210 BHN develops a hardness of over 400 BHN, and to an appreciable depth, after a short period of service. The steel can be welded, using suitably modified techniques, by either the thermit or flash butt welding processes.

14% Mn rails are used traditionally in railway points and crossings and in other situations where the extended service life justifies their higher cost compared with pearlitic grades.

At one time, it was postulated that the high rate of work hardening in this steel (also known as Hadfield's Manganese Steel) was due to the formation of strain-induced martensite. However, it is now known that the hardening effect is associated with the formation of stacking faults in the austenitic structure and strain-induced martensite will only form in decarburized material or in steels of a lower alloy content.

Micro-alloy forging steels

Up until the late 1940s, the engineering steels that were used for automotive engine and transmission parts were based largely on compositions containing substantial amounts of nickel and molybdenum. The philosophy that prevailed was that these components were subjected to arduous service conditions that required high levels of strength and toughness and that this combination of properties was best achieved in Ni–Mo grades. However, during the 1950s, there was the realization that many of the steels were over-alloyed with regard to the hardenability requirements of the components and that the specified levels of strength could be achieved by steels of leaner composition. In the 1960s, the emerging technology of fracture mechanics provided greater knowledge on the level of toughness required in engineering components and indicated that satisfactory performance could be provided by steel compositions which gave lower impact energy values than the traditional Ni–Mo grades. These factors, coupled with major advances in heat treatment technology, led the way to the gradual substitution of the Ni–Mo grades by cheaper steels involving additions of manganese, chromium and boron.

By the 1970s, the opportunities for alloy reduction and substitution had largely been exhausted but competition in the automotive industry maintained the impetus for further cost reduction. Attention therefore turned to potential savings in manufacturing costs and particularly in the area of heat treatment. Traditionally, components such as crankshafts and connecting rods are cooled to room temperature after the forging operation, only to be reheated to a temperature of about 850°C prior to oil quenching. Tempering at 550–650°C then produces tensile strengths in the range 800–1100 N/mm². However, in the mid-1970s, German manufacturers demonstrated that these strength levels could be produced in a micro-alloy, medium-carbon steel (49 MnVS3) after air cooling from the forging operation, thus eliminating the need for heat treatment. Since that time, major effort has been devoted to the development of micro-alloy forging steels in Europe and Japan and these steels have gradually been introduced as substitutes for quenched and tempered steels in some automotive components.

Metallurgical considerations

As indicated in earlier chapters, niobium, titanium and vanadium are used as micro-alloying elements in low-carbon steels, although high soaking temperatures must be employed in order to achieve substantial solution of Nb(CN), TiC and TiN. However, vanadium has a high solubility in austenite, regardless of the carbon content, and is therefore the most suitable micro-alloying element for medium-carbon steels. On cooling from the solution treatment temperature, vanadium carbonitride precipitates in both the proeutectoid ferrite and the ferrite lamellae of the pearlite. The physical metallurgy of these steels has been reviewed by Gladman.[46]

As illustrated in Figure 4.43, the tensile properties of these grades increase progressively with vanadium content and, depending upon the levels of strength required, vanadium contents in the range 0.05–0.2% are employed. The level of precipitation strengthening is also influenced by the nitrogen content and Lagnenborg[47] has shown that the tensile strength of these steels can be expressed as a function of $(V + 5 \times N)\%$. Nitrogen levels of up to 0.02% are therefore incorporated in the steels in order to intensify the strengthening effect.

One of the disadvantages of these micro-alloy steels is that they display significantly lower levels of toughness than the traditional quenched and tempered martensitic grades and this has inhibited their large-scale commercial exploitation. The low impact strength is related to the coarse pearlitic structure but this effect is exacerbated by precipitation strengthening. Whereas this problem has been overcome in structural steel plates with the use of low-temperature finishing (controlled rolling), there is little scope for the adoption of this practice in drop-forging operations due to the metal flow/die filling problems that occur at low forging temperatures.

The impact strength of these grades can be improved by lowering the carbon content and compensating for the loss in strength by increasing the manganese, vanadium and nitrogen contents. Experience in Sweden and Germany has

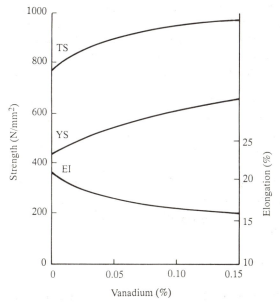

Figure 4.43 *Effect of vanadium on the tensile properties of air-cooled 0.45% C 0.9% Mn steel*

shown that an improvement in toughness can also be obtained by increasing the silicon content of the steels. However, attention has also been given to the potential of grain-refining additions of titanium in pursuit of higher impact strength.

As indicated earlier, titanium has a low solubility in medium-carbon steels but TiN is even less soluble than TiC. Particles of TiN are therefore present at the normal soaking temperature for forging, namely 1150°C, and will refine the austenite grains provided the particles are present as a fine dispersion. According to Gladman,[46] this is achieved by restricting the titanium content to below the stoichiometric level required for reaction with nitrogen in TiN and the growth of particles is also minimized by rapid solidification from the liquid state. In practice, the titanium additions are restricted to levels of about 0.01% and the need for rapid solidification is generally satisfied by continuous casting as opposed to ingot casting.

Japanese steelmakers have expressed concern that the formation of TiN for grain refinement can reduce the level of soluble nitrogen that is available for precipitation strengthening by V(CN). However, this problem can be overcome by adjusting the nitrogen content such that the free nitrogen (total nitrogen minus nitrogen as TiN) exceeds 0.006%.

More recently, it has been reported that improved levels of toughness can be obtained in medium-carbon, micro-alloy steels by the generation of bainitic structures. However, according to Naylor[48] the benefits of this development are not yet clear-cut and the need for alloy additions to achieve a bainitic structure may detract from the viability of the approach.

Commercial exploitation

Korchynsky and Paules[49] have reviewed the various grades of micro-alloy forging steels that are produced in Europe and Japan and their lists of compositions and associated tensile properties are shown in Tables 4.26 and 4.27. Heading the list is the German grade 49MnVS3, the first medium-carbon, micro-alloy steel to be used commercially for air-cooled automotive forgings. Like the Swedish Volvo grade V2906, 49 MnVS3 has a relatively low manganese content which restricts the amount of pearlite in the microstructure, and the tensile strengths of these grades are towards the bottom of the range. In the UK, substantial effort has been devoted to the development and commercial evaluation of the VANARD range which is based on the following composition range:

0.3–0.5% C, 0.15–0.35% Si, 1–1.5% Mn, 0.05–0.2% V

Within this composition range, increasing levels of carbon, manganese and vanadium are used to provide tensile strengths in the range 850–1100 N/mm².

Some of the quenched and tempered, alloy steels that can be replaced by the VANARD grades are shownin Table 4.28.

Table 4.26 *Chemical composition of micro-alloy forging steels*

Country	Grade	C	Si	Mn	S	V	Other
Germany	49MnVS3	0.44–0.5	0.6 max.	0.7–1	0.04–0.07	0.08–0.13	
Gr. Britain	BS970-280M01	0.3–0.5	0.15–0.35	0.6–1.5	0.045–0.06	0.08–0.2	
Gr. Britain (UES)	VANARD	0.3–0.5	0.15–0.35	1–1.5	0.1 max.	0.05–0.2	0.1 Cr
Gr. Britain (UES)	VANARD 850	0.36	0.17	1.25	0.04	0.09	0.15 Cr
Gr. Britain (UES)	VANARD 1000	0.43	0.35	1.25	0.06	0.09	0.04 Mo max.
Gr. Britain (Austin-Rover)	CMV 925	0.37–0.42	0.15–0.35	1.1–1.3	0.06–0.08	0.08–0.11	Cr+Cu+Ni=0.5 max.
Finland (OVAKO)	IVA 1000	0.47	0.5	1.1	0.05	0.13	0.5 Cr
Sweden (Volvo)	V-2906	0.43–0.47	0.15–0.4	0.6–0.8	0.04–0.06	0.07–0.1	0.2 Cr max.
Germany	44MnSiVS6	0.42–0.47	0.5–0.8	1.3–1.6	0.02–0.035	0.1–0.15	Ti optional
Germany	38MnSiVS6	0.35–0.4	0.5–0.8	1.2–1.5	0.04–0.07	0.08–0.13	Ti optional
Germany	27MnSiVS6	0.25–0.3	0.5–0.8	1.3–1.6	0.03–0.05	0.08–0.13	Ti optional
Japan (Mitsubishi-NKK)		0.32	0.25	1.45		0.06	0.01 Ti; 0.01–0.016 N

After Korchynsky and Paules.[49]

Table 4.27 *Tensile properties and hardness of micro-alloy forging steels*

Country	Grade	UTS (MPa)	YS (MPa)	El (% min.)	RA (% min.)	BHN
Germany	49MnVS3	750–900	450 min.	8	20	
Gr. Britain	BS970-280M01	780–1080	540–650	18/8	20	
Gr. Britain (UES)	VANARD 850	770–930	540 min.	18	20	237–277
Gr. Britain (UES)	VANARD 1000	930–1080	650 min.	12	15	269–331
Gr. Britain (Austin-Rover)	CMV 925	850–1000	560	12	15	248–302
Finland (OVAKO)	IVA 1000	1025	750	10	20	290
Sweden (Volvo)	V-2906: <90 mm	750–900	500 min.	12		230–275
	<50 mm	800–950	520 min.	15		245–290
Germany	44MnSiVS6	950–1100	600 min.	10	20	
Germany	38MnSiVS6	820–1000	550 min.	12	25	
Germany	27MnSiVS6	800–950	500 min.	14	30	
Japan (Mitsubishi-NKK)		720–800	470–550			

After Korchynsky and Paules.[49]

Table 4.28

Grade	Steel replaced
VANARD 850	0.35% C, 1.5% Mn (216M36)
VANARD 925	0.35% C, 1.5% Mn, 0.25% Mo (605M36) 0.4% C, 0.8% Mn, 1% Cr (530M40)
VANARD 1000	0.35% C, 1.5% Mn, 0.25% Mo (605M36) 0.4% C, 0.8% Mn, 1% Cr, 0.3% Mo (709M40)
VANARD 1100	0.4% C, 0.8% Mn, 1% Cr, 0.3% Mo (709M40)

Whereas the standard VANARD grades are generally made to a sulphur specification of 0.05% max., variants are produced with sulphur contents of the order of 0.08% for improved machinability. Other steels listed in Table 4.26 also specify enhanced sulphur levels. However, irrespective of the sulphur content, it is generally claimed that micro-alloy forging steels offer better machinability than traditional martensitic grades due to easier crack propagation in the predominantly pearlitic structure.

In addition to compositional effects, the mechanical properties of these steels are controlled by the soaking temperature, hot-working schedule and cooling rate to ambient temperature. The cooling rate from the finishing temperature is important since this controls the transformation temperature and the precipitation of V(CN). In conventional forging steels, the components may be placed in a bin after the forging operation and the slow cooling rates encountered in this situation would cause the precipitates to overage, resulting in a substantial loss in strength. Simple conveyer systems have therefore been introduced which enable the micro-alloy steel forgings to cool freely in air or else the cooling rate is enhanced with fan cooling.

The automotive components that are being produced in medium-carbon, micro-alloy steels include crankshafts, connecting rods, steering knuckles, axle beams and tension rods. Various manufacturers, including the Rover Group, have claimed that very substantial cost savings have been achieved by the adoption of these grades due to:

1. The lower cost of micro-alloy steels compared with the alloy grades that they replace.
2. The elimination of heat treatment costs.
3. The improved machining characteristics compared with traditional grades.

However, car makers in the UK have been reluctant to use these steels for safety-critical components due to their low impact properties and the rate of acceptance has been particularly slow in North America where manufacturers face greater threats of litigation on improper application/product liability. Therefore the further exploitation of these steels is very dependent on the development of improved toughness via controlled processing or grain-refinement techniques.

Controlled processed bars

There is a limited demand for normalized or quenched and tempered bar products that can be machined directly to the finished component form. Given that these heat treatments are expensive, there is obviously an incentive to develop the required mechanical properties in the as-rolled condition, i.e. the incentive is similar to that described in the previous section for the elimination of heat treatment in automotive forgings.

Normalized steels

Normalizing is applied to bar products in order to refine the grain size and facilitate subsequent processing operations such as machining or cold forging. The steel is reheated to just above the reaustenitization temperature (Ac_3) to achieve a fine austenite grain size and the material is then allowed to cool freely in air. In order to simulate such a structure in the 'as-rolled' condition, three options can be considered:

1. Lower billet reheating temperatures.
2. Lower bar finishing temperatures.
3. Faster cooling rates after rolling.

In rolling mills that do not have inter-stand cooling or delayed rolling facilities, options (1) and (2) are interdependent and both are aimed at the refinement of the austenite grain size. Whereas the austenite grain size will affect the pearlite content and ferrite grain size, these microstructural parameters are also influenced significantly by the rate of cooling from the finishing temperature.

Japanese steelmakers have introduced the term *normalize-free* for controlled rolled products that involve billet reheating temperatures of around 1050°C and finishing at or below 900°C in order to simulate the properties of normalized bars. Whereas such finishing temperatures can be achieved by introducing delays into the rolling schedule, economic rates of production can only be sustained with the adoption of inter-stand cooling.

Quenched and tempered steels

As indicated previously, the tensile properties of quenched and tempered alloy steel forgings can be reproduced in air-cooled, micro-alloy steels but the latter tend to produce inferior impact properties. However, rolling offers a greater opportunity than forging for low-temperature finishing and therefore a greater potential for improved toughness via grain refinement.

The relative effects of vanadium and niobium on strength–toughness relationships are shown in Figure 4.44. These additions were made to base compositions containing 0.33% C but with varying manganese. Under normal rolling conditions (reheat 1200°C, finish 1150°C), the vanadium addition produces the better impact properties (lower FATT) and the beneficial effect appears to increase slightly with increasing strength. Controlled rolling (reheat 1100°C,

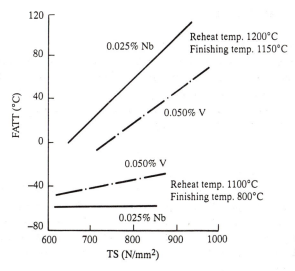

Figure 4.44 *Strength–toughness relationships in 0.33% C, micro-alloy steels*

finish 800°C) produces a major improvement in toughness in both types of steel but the niobium steel displays the better combination of properties, maintaining an FATT of about $-60°C$ over the range of tensile strength from 625 to 850 N/mm². The better performance of the niobium steel under controlled rolling conditions is probably related to the fact that niobium has a greater effect than vanadium in suppressing austenite recrystallization during hot rolling which leads to the production of a finer grain size. However, in situations where controlled rolling cannot be employed, then vanadium steels offer the more attractive properties.

High-carbon wire rod

Although tensile strengths > 2000 N/mm² are normally associated with lightly tempered martensites, or with maraging grades, these strength levels can also be achieved very readily in wire products by cold-drawing rods with a fine pearlitic microstructure. The demand for such products in the UK amounts to about 230 000 tonnes/annum and some of the more important applications are listed below:

- Wire ropes
- Prestressed concrete wire
- Tyre cord reinforcement
- Bridge suspension cables
- High-pressure hose reinforcement
- Helical springs (bedding and seating)
- Core wire for electrical conductor cables
- Piano strings

Rod rolling and conditioning

All wire is produced from hot-rolled rod and in the context of *high-carbon rod* this involves steels with carbon contents in the range 0.5–0.9%. Depending upon the carbon content, this results in the formation of a mixed ferrite–pearlite or a completely pearlitic microstructure. However, natural cooling in air from a high finishing temperature results in the generation of coarse pearlite which is unsuitable for severe cold-drawing operations. Traditionally, high-carbon rod is reheated to a temperature just above Ac_3, in order to reaustenitize the material, and then quenched into a lead bath at 450–500°C. The steel is therefore allowed to transform isothermally at a relatively low temperature to form a fine lamellar pearlite. Such a treatment is termed *patenting* and develops a high-strength structure which is also capable of extensive cold drawing. Patenting is still employed to a limited extent but most high-carbon steels are now drawn directly from as-rolled rod which has been subjected to an in-line cooling operation at the end of rod rolling. This produces a microstructure which is similar to that developed by the costly patenting process, but, on average, controlled cooled rod has a tensile strength which is about 100 N/mm² lower than that obtained on patenting.[50]

As indicated in the previous section, the strength of pearlitic steels is influenced very markedly by the carbon content, which controls the amount of pearlite in the structure, and also by the pearlite interlamellar spacing (S_o). However, for a given carbon content, the volume fraction of pearlite can be increased and the interlamellar spacing refined by depressing the austenite to pearlite transformation temperature. This can be achieved by:

1. Increasing the prior austenite grain size.
2. Increasing the rate of cooling from rod rolling.
3. Adding alloying elements such as chromium and manganese.

Jaiswal and McIvor[50] developed the following equation to illustrate the effects of cooling rate and composition on the tensile strength of controlled cooled, plain carbon steel rod:

$$\text{Tensile strength (N/mm}^2) = [267 \, (\log \text{CR}) - 293] + 1029 \, (\% \text{ C}) \\ + 152 \, (\% \text{ Si}) + 210 \, (\% \text{ Mn}) \\ + 442 \, (\% \text{ P})^{\frac{1}{2}} + 5244 \, (\% \text{ N}_f)$$

This equation relates to steels with a prior austenite grain size of ASTM 7 and CR is the cooling rate in °C/s at 700°C, i.e. before the start of transformation. Although nitrogen has the largest strengthening coefficient, carbon provides a much bigger contribution to the strength of these materials because it is present in very much higher concentrations.

Wire drawing

Prior to wire drawing, the hot-rolled rods must be cleaned in order to remove scale. This may be carried out mechanically by grit blasting or by subjecting the rods to a series of bending and twisting operations. However, for the more critical applications, acid pickling is employed followed by a neutralizing wash. The cleaned rod is then coated with lime or zinc phosphate for *dry drawing* or with a layer of copper or brass for *wet drawing*.

As illustrated in Figure 4.45, pearlitic steels work harden very rapidly during wire drawing to develop tensile strengths well in excess of 2000 N/mm^2.

Micro-alloy high-carbon rod

Jaiswal and McIvor[50] have described the use of micro-alloy additions in high-carbon rod in order to compensate for the lower tensile strengths that are achieved in controlled cooled rod compared with patented material. These authors show that additional strength can be achieved by the following mechanisms:

1. Refinement of the pearlite interlamellar spacing with the addition of chromium.
2. Solid solution strengthening using higher silicon contents.
3. Precipitation strengthening by the addition of vanadium.

However, micro-alloying increases the tendency for martensite formation which has a particularly damaging effect on the wire-drawing characteristics. The martensite-forming potential of various elements was therefore determined in controlled cooling experiments and the critical cooling rate at which martensite first appeared in the pearlitic matrix was determined. With a prior

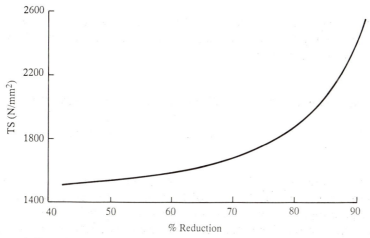

Figure 4.45 *Effect of cold drawing on the strength of high-carbon pearlitic rod*

austenite grain size of ASTM 7, the following critical cooling rate (CCR) relationship was derived:

$$CCR\ (Ks^{-1}) = 97 - (\%\ Si) - 70\ (\%\ Mn) - 50\ (\%\ Cr) - 224\ (\%\ P)$$

This equation indicates that manganese is potentially more damaging than chromium but it was also shown that chromium was more effective in producing a finer interlamellar spacing. Whereas silicon had relatively little effect in promoting the formation of martensite, it was also shown to be less effective than chromium in refining the interlamellar spacing. Jaiswal and McIvor therefore propose that silicon is used in combination with chromium, or possibly vanadium, rather than as a single addition.

Given the importance of avoiding the formation of martensite, Jaiswal and McIvor recommend that chromium is better used in large-diameter rods where the cooling rate is relatively slow and therefore less likely to produce martensite. For small-diameter rods, vanadium is recommended.

In a base steel containing 0.85% C and 0.7% Mn, the above authors state that the maximum levels of strength shown in Table 4.29 can be obtained by micro-alloying.

Table 4.29

Additive	Rod diameter	TS (N/mm^2)
0.8% Si, 0.25% Cr	Large	1330
0.07% V	Small	1300

References

1. Bain, E.C. and Davenport, E.S. *Trans. AIME*, **90**, 117 (1930).
2. Cias, W.W. *Phase Transformation Kinetics and Hardenability of Medium Carbon Alloy Steels*, Climax Molybdenum Co., Greenwich, Connecticut.
3. Atkins, M. *Atlas of Continuous Cooling Transformation Diagrams*, British Steel.
4. Siebert, C.A., Doane, D.V. and Breen, D.H. *The Hardenability of Steels*, ASM, Metals Park, Ohio (1977).
5. Grossman, M.A. *Elements of Hardenability*, ASM, Cleveland (1952).
6. Llewellyn, D.T. and Cook, W.T. *Metals Technology*, December, 517 (1974).
7. deRetana, A.F. and Doane, D.V. *Metal Progress*, September, **100**, 105 (1971).
8. Gladman, T. Private communication.
9. Ueno, M. and Inoue, J. *Trans. ISI Japan*, **13** (3), 210 (1973).
10. Kapadia, B.M., Brown, R.M. and Murphy, W.J. *Trans. Met. Soc. AIME*, **242**, 1689 (1968).

11. Smallman, R.E. *Modern Physical Metallurgy* (Fourth Edition), Butter-worths.
12. Grange, R.H. and Baughman, R.W. *Trans ASM*, **48**, 165 (1956).
13. Parrish, G. and Harper, G.S. *Production Gas Carburising*, Pergamon Press.
14. Murray, J.D. *Auto Engineer*, **55**, 186 (1965).
15. Llewellyn, D.T. and Cook, W.T. *Metals Technology*, May, 265 (1977).
16. Wannell, P.H., Blank, J.R. and Naylor, D.J. In *Proc. International Symposium on the Influence of Metallurgy on the Machinability of Steel*, September, ISIJ/ASM, Tokyo (1977).
17. Pickett, M.L. Cristinacce, M. and Naylor, D.J. In *Proc. High Productivity Machining, Materials and Processing*, May, ASM, New Orleans (1985).
18. Irani, R.S. *Metals and Materials*, June, 333 (1987).
19. Naylor, D.J. In *Proc. Integrity of Gas Cylinders: Materials Technology*, NPL, **11** (1985).
20. Oldfield, F.K. In *Proc. Integrity of Gas Cylinders: Materials Technology*, NPL, **1** (1985).
21. Harris, D., Priest, A., Davenport, J., McIntyre, P., Almond, E.A. and Roebuck, B. In *Proc. Integrity of Gas Cylinders: Materials Technology*, NPL, **69** (1985).
22. Zaretsky, E.V. *Effect of Steel Manufacturing Processes on the Quality of Bearing Steels* (ed. Hoo, J.C.C.), ASTM Special Technical Report, **5**, 981 (1988).
23. Bamberger, E.N. In *Proc. Tribology in the 80's* (ed. Loomis, W.F.), NASA CP-2300, Vol. 2, National Aeronautics and Space Administration, Washington, DC, 773 (1983).
24. Johnson, R.F. and Sewell, J. *JISI*, December, 414 (1960).
25. Brooksbank, D. and Andrews, K.W. *JISI*, **210**, 246 (1972).
26. Hampshire, J.M. and King, E. *Effect of Steel Manufacturing Processes on the Quality of Bearing Steels* (ed. Hoo, J.C.C.) ASTM Special Technical Report 981, 61 (1988).
27. Davies, I.G. Clarke, M.A. and Dulieu, D. *Effect of Steel Manufacturing Processes on the Quality of Bearing Steels*, ASTM Special Technical Report 981, 375 (1988).
28. Hoyle, G. *High Speed Steels*, Butterworth (1989).
29. *Source Book on Maraging Steels* (ed. Decker, R.F.), ASM.
30. Spitzig, W.A. In *Source Book on Maraging Steels* (ed. Decker, R.F.), 299.
31. Decker, R.F., Eash, J.T. and Goldman, A.J. In *Source Book on Maraging Steels* (ed. Decker, R.F.), 1.
32. *INCO Databook 1976*, International Nickel Co.
33. Magnee, A., Drapier, J.M., Dumont, J., Coutsouradis, D. and Hadbraken, L. *Cobalt-Containing High Strength Steels*, Cobalt Information Centre, Brussels (1974).
34. Collier, J.G. and Gemmill, M.G. *Metals and Materials*, April, 198 (1986).
35. Wyatt, L.M. *Materials of Construction for Steam Power Plant*, Applied Science Publishers (1976).
36. Reynolds, P.E., Barron, J.M. and Allen, G.B. *The Metallurgist and Materials Technologist*, July, 359 (1978).

37. Viswanathan, R. and Jaffee, R.I., *Trans. ASME*, **105**, October, 286 (1983).
38. Gemmill, M.G. *Metals and Materials*, December, 759 (1985).
39. Boyle, C.J., Curran, R.M., DeForrest, D.R. and Newhouse, D.L. *Proc. ASTM*, **62**, 1156 (1962).
40. Everson, H., Orr, J. and Dulieu, D. In *Proc International Conference on Advances in Material Technology for Fossil Power Plants* (eds Viswanathan, R. and Jaffee, R.I.) (Chicago, 1987), ASM International.
41. Pickering, F.B. and Gladman, T. *ISI Special Report 81* (1961).
42. Gladman, T., McIvor, I.D. and Pickering, F.B. *JISI*, **210**, 916 (1972).
43. Preston, R.R. *Steelresearch 87–88*, British Steel, 57.
44. Hodgson, W.H. and Preston, R.R. *CIM Bulletin*, October, 95 (1988).
45. British Steel *Track Products Brochure*, British Steel.
46. Gladman, T. *Ironmaking and Steelmaking*, **16**, No. 4, 241 (1989).
47. Lagnenborg, R. In *Proc. Fundamentals of Microalloying Forging Steels* (eds Krauss, G. and Banerji, S.K.), TMS of AIME, p. 39 (1987).
48. Naylor, D.J. *Ironmaking and Steelmaking*, **16**, No. 4 (1989).
49. Korchynsky, M. and Paules, J.R. *Microalloyed Forging Steels – A State of the Art Review*, SAE 890801 (1989).
50. Jaiswal, S. and McIvor, I.D. *Ironmaking and Steelmaking*, **16**, No. 1, 49 (1989).

5 *Stainless steels*

Overview

As chromium is added to steels, the corrosion resistance increases progressively due to the formation of a thin protective film of Cr_2O_3, the so-called *passive layer*. With the addition of about 12% Cr, steels have good resistance to atmospheric corrosion and the popular convention is that this is the minimum level of chromium that must be incorporated in an iron-based material before it can be designated a *stainless steel*. However, of all steel types, the stainless grades are the most diverse and complex in terms of composition, microstructure and mechanical properties. Given this situation, it is not surprising that stainless steels have found a very wide range of application, ranging from the chemical, pharmaceutical and power generation industries on the one hand to less aggressive situations in architecture, domestic appliances and street furniture on the other.

By the late 1800s, iron–chromium alloys were in use throughout the world but without the realization of their potential as corrosion-resistant materials. Harry Brearley, a Sheffield metallurgist, is credited with the discovery of martensitic stainless steels in 1913 when working on the development of improved rifle barrel steels. He found that a steel containing about 0.3% C and 13% Cr was difficult to etch and also remained free from rust in a laboratory environment. Such a steel formed the basis of the cutlery industry in Sheffield and as Type 420 is still used for this purpose to the present day.

During the same period, researchers in Germany were responding to pressures for improved steels for the chemical industry. Up until that time, steels containing high levels of nickel were in use as tarnish-resistant materials but had inadequate resistance to corrosion. Two Krupp employees, Benno Strauss and Eduard Maurer, are credited with the discovery of Cr–Ni austenitic stainless steels and patents on these materials were registered in 1912. However, workers in France and the United States are also cited as independent discoverers of these steels.

During the 1920s and 1930s, rapid developments took place which led to the introduction of most of the popular grades that are still in use today, such as Type 302 (18% Cr, 8% Ni), Type 316 (18% Cr, 12% Ni, 2.5% Mo), Type 410 (12% Cr) and Type 430 (17% Cr). However, even in the 1950s, stainless steels were still regarded as semi-precious metals and were priced accordingly. Up until the 1960s, these steels were still produced in small electric arc furnaces, sometimes of less than 10 tonnes capacity. The process was carried out in a single stage, involving the melting of scrap, nickel and ferro-chrome, with production times in excess of $3\frac{1}{2}$ hours. However, substantial gains were achieved with the installation of larger furnaces with capacities greater than 100 tonnes and the introduction of oxygen refining techniques also increased productivity very substantially. Since the early 1970s, the production of stainless steels has been based on a two-stage process, the first employing a conventional electric arc furnace for the rapid melting of scrap and ferro-alloys but using cheap, high-carbon ferro-chrome as the main source of chromium units. The high-carbon melt is then refined in a second stage, using either an argon–oxygen decarburizer (AOD) or by blowing with oxygen under vacuum (VOD). The AOD process is now employed for over 80% of the world's production of stainless steel and produces 100 tonnes of material in less than one hour. However, in addition to achieving faster production rates, the intimate mixing with special slags results in very efficient desulphurization. Other benefits also accrue from the facility to produce carbon contents of less than 0.01% and hydrogen levels of 2–3 ppm.

Substantial cost savings were also achieved with the adoption of continuous casting in place of ingot casting and these overall gains in production have led to a significant cheapening of stainless steel relative to two of its main competitors, namely plastics and aluminium. For the future, there is the prospect of the direct introduction of cheap chromium ores and their reduction by coal in a converter which would lead to further cost reduction.

In terms of product innovation, perhaps the greatest benefits have been obtained from relatively simple changes, such as the introduction of stainless grades with low carbon contents, i.e. below 0.03% C. This modification has virtually eliminated the risk of intergranular corrosion in unstabilized austenitic grades and has also improved the corrosion performance and ductility of ferritic grades. However, steel users have been reluctant to take advantage of higher strength austenitic steels, such as those based on 0.2% N, which can lead to significant cost reductions through the use of reduced thicknesses in pipework and pressure vessels. This contrasts sharply with the situation outlined earlier in Chapter 3 where the micro-alloy grades are now used extensively in place of plain carbon steel. When these high-nitrogen stainless grades were introduced in the UK in the mid-1960s, their high proof strength values could not be used to full advantage because of limitations imposed by the design codes of the day. Welding problems were also encountered due to the fact that the high nitrogen content led to the formation of a fully austenitic weld metal and susceptibility to solidification cracking. However, these problems have now been resolved and therefore there is the prospect of greater utilization of these materials in the future.

The 1970s saw the introduction of the *low interstitial ferritic* grades, with combined carbon and nitrogen contents of less than 200 ppm. These steels are based on compositions such as 18% Cr 2% Mo and 26% Cr 1% Mo and offered the prospect of being a cheaper alternative to an austenitic grade such as Type 316 (18% Cr, 12% Ni, 2.5% Mo). However, whereas the low interstitial grades exhibit good corrosion resistance, particularly with regard to chloride-induced stress corrosion, they tend to retain the problem of conventional ferritic steels in relation to grain coarsening and loss of toughness after welding. On the other hand, high-alloy steels involving duplex austenite plus ferrite microstructures are now gaining acceptance, because of their higher strength and better resistance to stress corrosion than conventional austenitic grades.

Whereas the consumption of bulk steel products is likely to remain fairly static, stainless steel is still very much in a growth market. This relates to the fact that stainless steel has managed to maintain its traditional image as a decorative material but is now also regarded as an engineering material for use in applications where structural integrity is more important that aesthetic appearance. On the basis of life-cycle costing, stainless steels are also proving to be attractive alternatives to mild steel in structures or components that require frequent painting and maintenance.

Composition–structure relationships

Iron–chromium alloys

The simplest stainless steels consist of iron–chromium alloys but in fact the binary iron–chromium system can give rise to a wide variety of microstructures with markedly different mechanical properties. The Fe–Cr equilibrium diagram is shown in Figure 5.1 and is characterized by two distinctive features, namely:

1. The presence of sigma phase at about 50% Cr.
2. The restricted austenite phase field, often called the *gamma-loop*.

Sigma phase is an *intermetallic compound*, which is hard and brittle and can be produced in alloys containing substantially less than 50% Cr. It also has an adverse effect on the corrosion resistance of stainless steels and therefore care should be taken to avoid extended exposure in the temperature range 750–820°C which favours its formation.

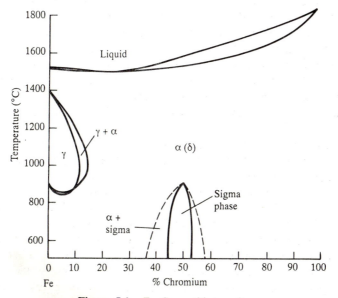

Figure 5.1 *Fe–Cr equilibrium diagram*

From a commercial standpoint, the area of the Fe–Cr diagram of greatest importance is that containing up to about 25% Cr, and a simplified illustration of that region for alloys containing 0.1% C is shown in Figure 5.2. Because chromium is a mild carbide former, many types of stainless steel are solution treated at temperatures significantly higher than those used for low-alloy steels in order to dissolve the chromium carbides. A solution treatment temperature of 1050°C is typical of a variety of stainless steel grades and this will be used as a reference temperature in relation to the microstructure at high temperature. As illustrated in Figure 5.2, 0.1% C steels can accommodate up to about 13.5% Cr at 1050°C and still remain austenitic with a face-centred-cubic structure. As the chromium content of the steels is increased within this range, the hardenability also increases very substantially such that large section sizes can be through-hardened to martensite on cooling to room temperature. For example, a steel containing 12% Cr and 0.12% C will form martensite at the centre of a 100 mm bar on air cooling from 1050°C and the limiting section can be increased to about 500 mm by oil quenching from this temperature. It should also be noted that the M_s–M_f transformation range is depressed significantly with large additions of chromium. However, for most commercial grades of 11–13% Cr steels, the transformation range is above room temperature and therefore the formation of retained austenite is not a major problem.

As the chromium content is increased above about 13%, a significant change takes place in the microstructure at 1050°C as the single-phase austenite region gives way to the *duplex* austenite plus ferrite phase field. Whereas the austenite in this region behaves in a similar way to that within the gamma-loop, i.e. it transforms to martensite on cooling to room temperature, the ferrite formed at

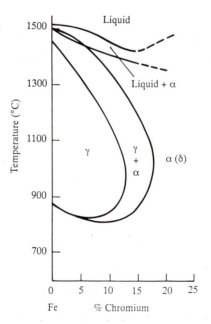

Figure 5.2 *Fe–Cr alloys containing 0.1% C*

high temperature undergoes no phase transformation. Although its structure is body-centred-cubic, the high-temperature phase is generally called delta ferrite (δ) in order to differentiate it from alpha ferrite (α), the transformation product from austenite. As illustrated in Figure 5.3, the delta ferrite content increases progressively with further additions of chromium and, in the presence of about 0.1% C, the material becomes completely ferritic with the addition of just over 18% Cr. Therefore, between about 13% and 18% Cr, the hardness of these steels is reduced as the microstructure changes progressively from 100% martensite to 100% delta ferrite. Larger additions of chromium have no further effect on the microstructure, although such materials become increasingly more susceptible to the formation of sigma phase.

Although the changes from $\gamma \rightarrow \gamma + \delta$ and $\gamma + \delta \rightarrow \delta$ occur at chromium levels of about 13.5% and 18.5% respectively, it must be emphasized that this refers to alloys containing a maximum of about 0.1% C and at a reference temperature of around 1050°C. Larger amounts of carbon or the addition of other alloying elements will have a major effect on the microstructure associated with particular levels of chromium. Additionally, for a given composition, an increase in the solution treatment temperature above 1050°C will also increase the amount of delta ferrite at the expense of austenite.

Iron–chromium–nickel alloys

Whereas chromium restricts the formation of austenite, nickel has the opposite effect and, as illustrated in Figure 5.4, the Fe–Ni equilibrium diagram displays an expanded austenite phase field. In the context of stainless steels, chromium is therefore termed a *ferrite former* and nickel an *austenite former*. Thus having created a substantially ferritic microstructure with a large addition of chromium, it is possible to reverse the process and re-establish an austenitic structure by adding a large amount of nickel to a high-chromium steel.

As indicated in Figure 5.3, a steel containing 17% Cr and 0.1% C will have a microstructure of about 65% delta ferrite–35% austenite at a solution treatment temperature of 1050°C. The various changes that then occur with the addition

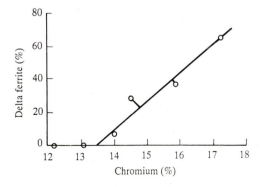

Figure 5.3 *Effect of chromium content on 0.1% carbon steels solution treated at 1050°C (After Irvine et al.[1])*

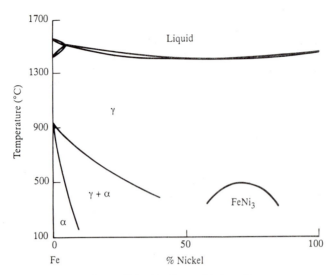

Figure 5.4   *Fe–Ni equilibrium diagram*

of nickel to a base steel of this composition are illustrated in Figure 5.5. This shows that the delta ferrite content is steadily reduced and at 1050°C the steels become fully austenitic with the addition of about 5% nickel. On cooling to room temperature, the austenite in these low-nickel steels transforms to martensite and therefore there is initially a progressive increase in hardness with the addition of nickel as martensite replaces delta ferrite. However, the addition of nickel also depresses the M_s–M_f transformation range and at nickel contents greater than about 4% the M_f temperature is depressed below room temperature. Further additions of nickel therefore lead to a decrease in hardness due to incomplete transformation to martensite and the formation of retained austenite. As indicated by the hardness data in Figure 5.5, refrigeration at −78°C causes the retained austenite to transform to martensite over a limited composition range until the M_s temperature coincides with the refrigeration temperature. In commercial 18% Cr 9% Ni austenitic stainless steels, the M_s has been depressed to very low temperatures and little transformation to martensite can be induced, even at the liquid nitrogen temperature of −196°C.

Other alloy additions

Whereas chromium and nickel are the principal alloying elements in stainless steels, other elements may be added for specific purposes and therefore consideration must be given to the effect of these elements on microstructure. Like chromium and nickel, these other alloying elements can be classed as ferrite or austenite formers and their behaviour is illustrated in Figure 5.6, which refers to a base steel containing 17% Cr and 4% Ni. Thus elements such as aluminium, vanadium, molybdenum, silicon and tungsten behave like chromium and promote the formation of delta ferrite. On the other hand, copper, manganese, cobalt, carbon and nitrogen have a similar effect to nickel

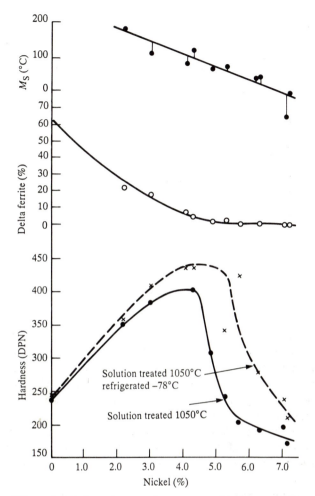

Figure 5.5 *Effect of nickel content on the structure, hardness and* M_s *temperature of 0.1% C 17% Cr steels (After Irvine et al.[1])*

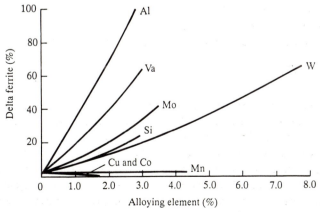

Figure 5.6 *Effect of various alloying elements on the structure of 17% Cr 4% Ni alloys (After Irvine et al.[1])*

and promote the formation of austenite. A guide to the potency of the various elements in their role as austenite or ferrite formers[1] is shown in Table 5.1.

Table 5.1

Element	Change in delta ferrite per 1.0 wt %	
N	− 200	
C	− 180	
Ni	− 10	
Co	− 6	Austenite
Cu	− 3	formers
Mn	− 1	
W	+ 8	
Si	+ 8	
Mo	+ 11	Ferrite
Cr	+ 15	formers
V	+ 19	
Al	+ 38	

Thus carbon and nitrogen are particularly powerful austenite formers and the latter is incorporated in certain grades of stainless steel, specifically for this purpose. Elements such as titanium and niobium are also ferrite formers in their own right but have an additional ferrite-promoting effect by virtue of the fact that they are also strong carbide and nitride formers and can therefore eliminate the austenite-forming effects of carbon and nitrogen.

Whereas alloying elements oppose each other in terms of austenite or ferrite formation at elevated temperatures, they act in a similar manner in depressing the martensite transformation range. Andrews[2] has derived the following formula for the calculation of M_s:

$$M_s \ (°C) = 539 - 423C - 30.4Mn - 17.7Ni - 12.1Cr - 7.5Mo$$

Therefore, in predicting the room temperature microstructure of stainless steels, consideration has to be given to two major effects:

1. The balance between austenite and ferrite formers which dictates the microstructure at elevated temperatures.
2. The overall alloy content which controls the M_s–M_f transformation range and the degree of transformation to martensite at ambient temperature.

A convenient but very approximate method of relating composition and microstructure in stainless steels is by means of the Schaeffler diagram which has been modified by Schneider.[3] This is illustrated in Figure 5.7, which indicates the structures produced in a wide range of compositions after rapid cooling from 1050°C. In this diagram, the elements that behave like chromium

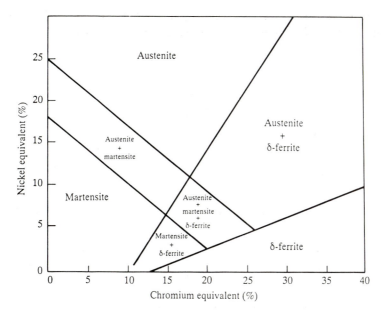

Figure 5.7   *Schaeffler diagram – modified (After Schneider³)*

in promoting the formation of ferrite are expressed in terms of a *chromium equivalent*:

$$\text{Cr equivalent} = (\text{Cr}) + (2\text{Si}) + (1.5\text{Mo}) + (5\text{V}) + (5.5\text{Al}) + (1.75\text{Nb}) + (1.5\text{Ti}) + (0.75\text{W})$$

In a similar manner, the austenite-forming elements are expressed in terms of a *nickel equivalent*:

$$\text{Ni equivalent} = (\text{Ni}) + (\text{Co}) + (0.5\text{Mn}) + (0.3\text{Cu}) + (25\text{N}) + (30\text{C})$$

all concentrations being expressed as weight percentages.

Commercial grades of stainless steels

From the foregoing remarks, it can be appreciated that stainless steels embrace a wide range of microstructures which are controlled by means of a complex relationship with composition. Although chromium may be the principal alloying element in a stainless steel, the level may give little indication of structure, and steels containing 17% Cr can be martensitic, ferritic or austenitic, depending on heat treatment and the presence of other elements. In discussing the characteristics of stainless steels, it is therefore more convenient to categorize these materials in terms of microstructure rather than composition.

Although most industrialized countries have developed their own national standards for stainless steels, these steels are referred to almost universally by means of the American Iron and Steel Institute (AISI) numbering system. Thus the martensitic stainless steels are classified as the 400 series but, rather confusingly, the 400 series also includes the ferritic grades of stainless steel. The more important grades of austenitic stainless steel are classified in the 300 series.

In the UK, the familiar specification for stainless steels is BS 1449: 1983 *Steel plate, sheet and strip Part 2: Stainless and heat resisting steels* and an extract from this specification is shown in Table 5.2. A compilation of comparative national standards in other countries, for the more important grades, is given in Table 5.3. However, work is now well advanced to produce a single European standard for stainless steel and it is anticipated that the following parts of a unified standard will be published towards the end of 1995:

EN 10088-1   Stainless steels – Part 1: List of stainless steels
EN 10088-2   Stainless steels – Part 2: Technical delivery conditions for sheet/plate and strip for general purposes
EN 10088-3   Stainless steels – Part 3: Technical delivery conditions for semi-finished products, bars, rods and sections for general purposes

These standards will therefore replace BS 1449: Part 2: 1993 but will not include pressure vessel and heat resisting grades which will feature in a separate specification at a later date.

It is understood that EN 10088 will include 20 ferritic grades, 20 martensitic grades, 37 austenitic grades and 6 duplex grades and will be more comprehensive than BS 1449: 1983 in terms of the range of steel grades, property data and available finishes.

Martensitic stainless steels

A fully martensitic structure can be developed in a stainless steel provided:

1. The balance of alloying elements produces a fully austenitic structure at the solution treatment temperature, e.g. 1050°C.
2. The M_s–M_f temperature range is above room temperature.

As indicated earlier, these conditions are met in the case of a 12% Cr 0.1% C steel and such a grade (Type 410) defines the lower bound of composition in the commercial range of martensitic stainless steels. However, given that the M_s temperature of a 12% Cr 0.1% C steel is substantially above room temperature, there is the facility to make further alloy additions and still maintain the martensitic transformation range above room temperature. Therefore the alloy content can be increased in order to obtain the following improvements in properties:

1. A higher martensitic hardness by means of an increased carbon content.

Table 5.2 British Standard 1449: 1983: Steel Plate, Sheet and Strip Part 2: Stainless and Heat Resisting Steels
Properties minima unless stated

Grade	Composition (maxima unless stated)									Condition	Tensile strength R_m N/mm²	Proof stress $R_{p0.2}$ N/mm²	Proof stress $R_{p1.0}$ N/mm²	Elongation[1] % min	Hardness[2] HV max
	C	Si	Mn	P	S	Cr	Mo	Ni	Others						
Austenitic Steels															
284S16	0.07	1.00	7.00/10.00	0.060	0.030	16.5/18.5	—	4.0/6.50	N 0.15/0.25	Softened	630	300	335	40	220
301S21	0.15	1.00	2.00	0.045	0.030	16.0/18.0	—	6.0/8.0	—	Softened	540	215	250	40	220
304S11	0.030	1.00	2.00	0.045	0.030	17.0/19.0	—	9.0/12.0	—	Softened	480	185	215	40	185
304S15	0.06	1.00	2.00	0.045	0.030	17.5/19.0	—	8.0/11.0	—	Softened	500	195	230	40	190
304S16	0.06	1.00	2.00	0.045	0.030	17.5/19.0	—	9.0/11.0	—	Softened	500	195	230	40	190
304S31	0.07	1.00	2.00	0.045	0.030	17.0/19.0	—	8.0/11.0	—	Softened	500	195	230	40	185
305S19	0.10	1.00	2.00	0.045	0.030	17.0/19.0	—	11.0/13.0	—	Softened	490	185	220	40	185
309S24	0.15	1.00	2.00	0.045	0.030	22.0/25.0	—	13.0/16.0	—	Softened	510	205	240	40	205
310S24	0.15	1.00	2.00	0.045	0.030	23.0/26.0	—	19.0/22.0	—	Softened	510	205	240	40	205
315S16	0.07	1.00	2.00	0.045	0.030	16.5/18.5	1.25/1.75	9.0/11.0	—	Softened	510	205	240	40	205
316S11	0.030	1.00	2.00	0.045	0.030	16.5/18.5	2.0/2.5	11.0/14.0	—	Softened	490	190	225	40	195
316S13	0.030	1.00	2.00	0.045	0.030	16.5/18.5	2.5/3.0	11.5/14.5	—	Softened	490	190	225	40	195
316S31	0.07	1.00	2.00	0.045	0.030	16.5/18.5	2.0/2.5	10.5/13.5	—	Softened	510	205	240	40	205
316S33	0.07	1.00	2.00	0.045	0.030	16.5/18.5	2.5/3.0	11.0/14.0	—	Softened	510	205	240	40	205
317S12	0.030	1.00	2.00	0.045	0.030	17.5/19.5	3.0/4.0	14.0/17.0	—	Softened	490	195	230	40	195
317S16	0.06	1.00	2.00	0.045	0.030	17.5/19.5	3.0/4.0	12.0/15.0	—	Softened	510	205	240	40	205
320S31	0.08	1.00	2.00	0.045	0.030	16.5/18.5	2.0/2.5	11.0/14.0	Ti 5C/0.80	Softened	510	210	245	40	205
320S33	0.08	1.00	2.00	0.045	0.030	16.5/18.5	2.5/3.0	11.5/14.5	Ti 5C/0.80	Softened	510	210	245	40	205
321S31	0.08	1.00	2.00	0.045	0.030	17.0/19.0	—	9.0/12.0	Ti 5C/0.80	Softened	500	200	235	35	200
347S31	0.08	1.00	2.00	0.045	0.030	17.0/19.0	—	9.0/12.0	Nb 10C/1.00	Softened	510	205	240	35	200

Grade	Composition (maxima unless stated)									Condition	Tensile strength R_m N/mm²	Proof stress $R_{p0.2}$ N/mm²	Elongation[1] % min	Hardness[2] HV max	
	C	Si	Mn	P	S	Cr	Mo	Ni	Others					Plate	Sheet/Strip
Ferritic Steels															
403S17	0.08	1.00	1.00	0.040	0.030	12.0/14.0	—	1.00	—	Softened	420	245	20	190	175
405S17	0.08	1.00	1.00	0.040	0.030	12.0/14.0	—	1.00	Al 0.10/0.30	Softened	420	245	20	190	175
409S19	0.08	1.00	1.00	0.040	0.030	10.5/12.5	—	1.00	Ti 6C/1.0	Softened	350	200	20	190	175
430S17	0.08	1.00	1.00	0.040	0.030	16.0/18.0	—	1.00	—	Softened	430	245	20	190	175
434S17	0.08	1.00	1.00	0.040	0.030	16.0/18.0	0.90/1.30	1.00	—	Softened	430	245	20	—	185
Martensitic Steels															
410S21	0.09/0.15	1.00	1.00	0.040	0.030	11.5/13.5	—	1.00	—	Softened	—	—	—	190	185
420S45	0.28/0.36	1.00	1.00	0.040	0.030	12.0/14.0	—	1.00	—	Softened	—	—	—	230	220

1. Elongation measured on flat test pieces using 50mm gauge length or $5.65\sqrt{S_o}$. For cylindrical test pieces, a gauge length of $5.65\sqrt{S_o}$ is used.
2. For items of adequate thickness the Brinell hardness test may be used, applying the same limiting HB hardness values as given for HV.

Table 5.3 *Comparative international stainless steel grades*

British	French	German	Italian	Japanese	Swedish	USA
301S21	Z12CN17.08	1.4310	X12CrNi 17 07	SUS301	14 23 31	301
304S31				SUS302	14 23 32	302
304S15 304S16	Z6CN18.09	1.4301	X5CrNi 18 10	SUS304	14 23 33	304
304S11	Z2CN18.10	1.4306	X2CrNi 18 11	SUS304L	14 23 52	304L
305S19	Z8CN18.12		X8CrNi 18 12	SUS305		305
309S24	Z15CN24.13		X16CrNi 23 14	SUS309		309
310S24	Z12CN25.20	1.4845	X22CrNi 25 20	SUS310S	14 23 61	310
315S16					14 23 40	
316S31 316S33	Z6CND17.11	1.4401 1.4436	X8CrNiMo 17 13	SUS316	14 23 47 14 23 43	316
316S11 316S13	Z2CND17.12	1.4404 1.4435	X2CrNiMo 17 12	SUS316L	14 23 48 14 23 53	316L
317S12	Z2CND19.15	1.4435	X2CrNiMo 18 16	SUS317L	14 23 67	317L
317S16		1.4436		SUS317		317
320S31 320S33	Z8CND17.12	1.4571 1.4573			14 23 66 14 23 50	
321S31	Z6CNT18.12	1.4541	X6CrNiTi 18 11	SUS321	14 23 37	321
347S31	Z6CNNb 18.11	1.4550	X6CrNiNb 18 11 X8CrNiNb 18 11	SUS347	14 23 38	347
403S17	Z6C13	1.4000	X6Cr 13	SUS403	14 23 01	403
405S17	Z6CA13	1.4002	Z6CrAl 13	SUS405		405
409S19		1.4512				409
430S17	Z8C17	1.4016	X8Cr 17	SUS430	14 23 20	430
434S17	Z8CD17.01	1.4113	X8CrMo 17	SUS434	14 23 25	434
410S21	Z12C13	1.4006 1.4024	X12Cr 13	SUS410	14 23 02	410
420S45	Z30C13		X30Cr 13	SUS420J2	14 23 04	420

After British Steel Stainless.[4]

2. Improved tempering resistance and toughness via balanced additions of molybdenum, vanadium and nickel.

Type 410 (12% Cr, 0.1% C) is probably the most popular grade in the martensitic series and is used in a wide variety of general engineering applications in both the wrought and cast condition. In Type 420, the carbon content is raised to about 0.3% to provide increased hardness and is typical of the steels used in cutlery. Type 416 and 416Se are free cutting versions of 12% Cr steel and contain additions of 0.15% S min. and 0.15% Se min. respectively in order to provide improved machining characteristics.

During the 1950s, a considerable amount of work was carried out on the development of high-strength 12% Cr steels, particularly in relation to the requirements of the power generation industry for improved steam turbine bolting and blading materials. Two publications from that period by Irvine *et al.*[5] and Irvine and Pickering[6] still serve as classic texts on the metallurgy of high-strength 12% Cr steels. In the solution-treated condition (1050°C AC), a 12% Cr 0.1% C steel develops a tensile strength of about 1300 N/mm² and must be tempered in order to achieve a good combination of strength and toughness. Depending on the strength requirement, 12% Cr steels are tempered at temperatures up to about 675°C and the above authors adopted the temperature–time parameter approach to demonstrate the effects of alloying elements on the tempering behaviour. An example is shown in Figure 5.8 for a 12% Cr 0.14% C steel, the tempering parameter being $T(20 + \log t) \times 10^{-3}$ where T is the temperature in °A and t the time in hours. Detailed electron metallographic studies revealed the following changes during the tempering of the steel:

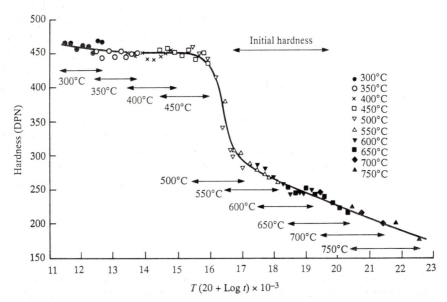

Figure 5.8 *Tempering curve for 0.14% C 12% Cr base steel (After Irvine et al.[5])*

1. In the solution-treated condition, the hardness was 455 HV and there was some evidence for the precipitation of small particles of Fe_3C within the martensitic structure, i.e. slight *autotempering*.
2. After tempering at 350°C, the amount of Fe_3C precipitation had increased, but at 450 HV the loss of hardness was still negligible.
3. Tempering at 450°C caused slight secondary hardening and this was associated with the precipitation of Cr_7C_3 and a small amount of M_2X (based on Cr_2C).
4. A major loss in hardness occurred at 500°C and this was associated with the precipitation of relatively large particles of chromium-rich $M_{23}C_6$ at the martensite plate and prior austenite grain boundaries.
5. Further softening occurred as the tempering temperature was increased and this was associated with the solution of Cr_7C_3 and the growth of $M_{23}C_6$ particles.

On tempering at 600°C, the steel had lost virtually all of the hardness associated with a martensitic structure and there was a particular need to improve the tempering resistance of 12% Cr turbine blading material at temperatures up to at least 650°C. This was concerned with a requirement for good impact properties but brazing operations for the attachment of wear-resistant shields also raised the temperature of the blades locally to a temperature of about 650°C.

As indicated in Chapter 4, additions of molybdenum and vanadium are very effective in improving the tempering resistance of low-alloy martensitic steels and the same is true for their 12% Cr counterparts. This is illustrated in Figure 5.9, which shows the effects of molybdenum and vanadium in a base steel containing 12% Cr, 2% Ni and 0.1% C. In either case, the addition of nickel is made to the steels to counteract the ferrite-forming potential of these elements. In the case of molybdenum (Figure 5.9(b)), a progressive increase in hardness is obtained on tempering at temperatures up to 500°C. This is due to an intensification of the secondary hardening reaction and a fine dispersion of M_2X precipitates persists at high tempering parameters. The addition of vanadium (Figure 5.9(c)) produces a similar effect and the stabilization of M_2X maintains a high level of hardness, even after tempering at 650°C.

Irvine and Pickering[6] also demonstrated the beneficial effects of carbon, nitrogen and niobium on the tempering resistance of 12% Cr steels. As illustrated in Figure 5.10, carbon and nitrogen increase the hardness throughout the tempering range and in the tempered 650°C 1 h condition, the 0.2% PS values of 12% Cr–Mo–V steels can be expressed by means of the following parameter:

$$0.2\% \text{ PS} = 710 + 772 \ (C + 2N) \ \text{N/mm}^2$$

Carbon and nitrogen also have the further benefit of being austenite formers and therefore do not require compensating additions to preserve the required martensitic structure.

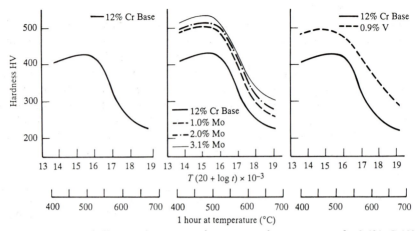

Figure 5.9   *Effect of alloying elements on the tapering characteristics of a 0.1% C 12% Cr steel (After Irvine and Pickering[6])*

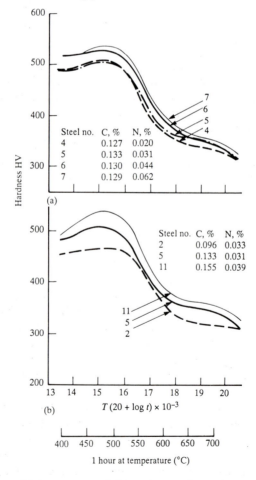

Figure 5.10   *Effect of (a) nitrogen and (b) carbon on the tempering characteristics of a 12% Cr–Mo–V steel (After Irvine and Pickering[6])*

The action of niobium in promoting tempering resistance is different to that of the other elements discussed above in that it intensifies the secondary hardening reaction by increasing the lattice parameter of the precipitate relative to that of the ferrite matrix. Thus the effect can be superimposed on other tempering retarding reactions, involving the stabilization of precipitates, in order to gain additional benefit. This is demonstrated in the tensile data given in Table 5.4 for 12% Cr 2.5% Ni 1.5% Mo 0.3% V steels, tempered 650°C 1 h. These data also indicate that nitrogen produces a powerful strengthening effect, albeit with a significant loss in toughness.

Table 5.4

Property	Base + 0.015% N	Base + 0.045% N	Base + 0.37% Nb, 0.013% N	Base + 0.33% Nb, 0.043% N
TS (N/mm²)	936	1055	981	1084
0.2% PS (N/mm²)	772	871	815	962
Elongation $4\sqrt{A}\%$	22.8	22.4	18.1	19.1
Charpy V – J at RT	94	47	104	52

As indicated earlier, the carbon content of 12% Cr steels can be increased to high levels in order to promote higher hardness levels but at the expense of toughness and weldability. Notable examples are cutlery steels containing 12% Cr and 0.3% C and stainless razor steels containing 12% Cr and 0.6% C. In the latter steel, substantial amounts of carbon remain out of solution as $M_{23}C_6$ carbides after solution treatment at 1050°C so that the full martensitic hardness associated with 0.6% C is not realized. However, the presence of carbides in the microstructure improves the abrasion resistance of such steels.

Ferritic stainless steels

Ferritic stainless steels can contain up to 30% Cr with additions of other elements such as molybdenum for improved *pitting corrosion* resistance and titanium or niobium for improved resistance to *intergranular corrosion*. These forms of corrosion will be described in a later section. However, the bulk of the requirement for ferritic stainless steels is satisfied by two major grades, namely Type 430 and 434. As shown in Table 5.2, Type 430S17 is a 17% Cr steel and at the normal solution treatment temperature of 950°C, this steel generally contains a proportion of austenite which transforms to martensite on cooling to room temperature. However, on tempering at 750°C, the martensite breaks down to ferrite and carbide, giving a microstructure which is essentially fully ferritic.

The corrosion resistance of stainless steels in chloride environments is improved substantially with the addition of molybdenum, and Type 434 (434S17) (17% Cr, 1% Mo) is the most common grade of this type within the ferritic range.

As illustrated later, the corrosion resistance of stainless steels can be seriously impaired by the precipitation of chromium carbides at the grain boundaries. One method of overcoming this problem is to add elements such as titanium and niobium which prevent the formation of chromium carbide. This gives rise to grades such as Type 430 Ti (17% Cr–Ti) and Type 430Nb (17% Cr–Nb).

As discussed earlier, 12% Cr steels are normally martensitic and, as such, tend to have poor forming and welding characteristics. However, the addition of a strong ferrite former to a 12% Cr base steel can produce a fully ferritic microstructure with a marked improvement in the cold-forming and welding behaviour. This effect is achieved in Type 409 (12% Cr–Ti), the titanium also eliminating the problem of chromium carbide formation as well as promoting the ferritic structure. This steel has found extensive application in automobile exhausts in place of plain carbon or aluminized steel.

Type 446 (25% Cr) is the highest chromium grade in the traditional range of ferritic stainless steels and provides the best corrosion and oxidation resistance. Whereas ferritic stainless steels generally possess low toughness, a further embrittling effect can be experienced in steels containing more than 12% Cr when heated to temperatures in the range 400–550°C. The most damaging effect occurs at a temperature of about 475°C and, for this reason, the effect is known as *475°C embrittlement*. The loss of toughness is due to the precipitation of a chromium-rich, α prime phase which becomes more pronounced with increase in chromium content. However, 475°C embrittlement can be removed by reheating to a temperature of about 600°C and cooling rapidly to room temperature.

Austenitic stainless steels

As illustrated in Table 5.2, most of the steels in the AISI 300 series of austenitic steels are based on 18% Cr but with relatively large additions of nickel in order to preserve the austenitic structure. However, as illustrated below, various compositional modifications are employed in order to improve the corrosion resistance of these steels.

Type 304 (18% Cr, 9% Ni) is the most popular grade in the series and is used in a wide variety of applications which require a good combination of corrosion resistance and formability. As discussed later, the stability and work-hardening rate of austenitic stainless steels are related to composition, the leaner alloy grades exhibiting the greater work hardening. Thus a steel such as Type 301 (17% Cr, 7% Ni) work hardens more rapidly than Type 304 (18% Cr, 9% Ni) and, for this reason, Type 301 is often used in applications calling for high abrasion resistance. Steels such as Type 316 (18% Cr, 12% Ni, 2.5% Mo) and 317 (18% Cr, 15% Ni, 3.5% Mo) have greater resistance to corrosion in chloride environments than Type 304 and represent austenitic counterparts of Type 434 (17% Cr, 1% Mo) discussed in the previous section.

Reference was made earlier to the corrosion problems experienced in stainless steels with the formation of chromium carbides. It was indicated that the problem can be overcome in ferritic stainless steels with the addition of titanium and niobium and the same is true in the austenitic grades. This gives rise to Types 321 (18% Cr, 10.5% Ni, Ti) and 347 (18% Cr, 11% Ni, Nb) and, because of their freedom from chromium carbide precipitation and intergranular attack, these grades are often referred to as *stabilized* stainless steels.

An alternative method of preventing the formation of chromium carbide in stainless steels is to reduce the carbon content to a low level. Thus Type 304L (0.03% C max.) is a lower carbon variety of Type 304 and is now used extensively in applications calling for resistance to intergranular attack in the welded condition. In a similar manner, Type 316L is the lower carbon version of Type 316.

Type 310 (25% Cr, 20% Ni) represents the most highly alloyed composition in the popular range of austenitic stainless steels and provides the greatest resistance to corrosion and oxidation.

During the early 1950s, there was a scarcity of nickel in the United States and this led to the development of austenitic stainless steels in which some of the nickel was replaced by alternative austenite-forming elements. The most successful steels of this type were Types 201 and 202, which have the mean compositions shown in Table 5.5.

Table 5.5

	$C\%$	$Si\%$	$Mn\%$	$Cr\%$	$Ni\%$	$N\%$
Type 201	0.1	0.5	6.5	17	4.5	0.25
Type 202	0.1	0.5	8.75	18	5	0.25

These steels were developed as alternatives to Types 301 (17% Cr, 7% Ni) and 302 (18% Cr, 8% Ni), in which reductions in the nickel content were compensated by large additions of manganese and nitrogen. The addition of 0.25% N also causes substantial solid solution strengthening and therefore the 200 series steels have high tensile properties. These steels were used extensively for the production of railway carriages in the United States but found little application in the UK. However, they are difficult to produce because of excessive refractory attack and problems in descaling due to their high manganese contents.

Nitrogen additions of about 0.2% are also made to standard grades such as Types 304 and 316 in order to generate high proof stress values. In the UK, these steels are marketed under the *Hyproof* tradename and provide 0.2 PS values of about 330 N/mm^2 compared with 250 N/mm^2 in the standard grades (typically 0.04% N).

The strength of austenitic steels can also be increased by warm working, i.e. by finishing rolling at temperatures below 950°C. As illustrated in Figure 5.11, this results in a major increase in the 0.2% proof stress of standard grades such as Type 304 and 321, but significantly higher levels of strength can be obtained by warm working compositions which are substantially solid solution strengthened with nitrogen. Steels of this type have found limited application in pressure vessels and also for high-strength concrete reinforcement.

Controlled transformation stainless steels

In the 1950s, interest was generated in an entirely new type of stainless steel that was austenitic in the as-delivered, solution-treated condition but which transformed to martensite by means of a simple heat treatment. Such materials offered the prospect of combining the good forming properties of traditional

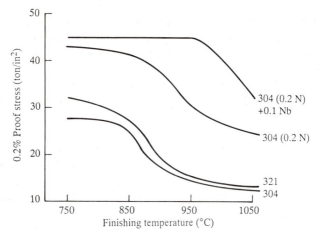

Figure 5.11 *Effect of finishing temperature on the properties of austenitic stainless steel (After McNeely and Llewellyn[7])*

austenitic stainless steels with the high strength of martensitic grades. The concept required that the composition of these steels was controlled to within fine limits such that the M_s–M_f temperature range was just below room temperature. Following cold-forming operations, the steel could then be transformed to martensite by refrigeration at a temperature such as $-78°C$ (solid CO_2). Such steels became known as *controlled transformation stainless steels* and their behaviour can be illustrated by reference to Figure 5.5. This indicates that a steel containing 0.1% C, 17% Cr and 5.5% Ni is essentially free of delta ferrite and has an M_s temperature near to ambient. In the solution-treated 1050°C AC condition, this composition has a hardness of about 220 HV, but on refrigeration at $-78°C$, the hardness increases to about 400 HV. This indicates that the steel has been substantially transformed to martensite, although Figure 5.5 shows that hardness values approaching 450 HV are obtained on complete transformation to martensite.

One of the problems encountered in the production of these steels was the very tight control of composition that was required in order to position the transformation range just below room temperature. If the alloy content was excessive, the transformation range was depressed to low temperatures such that transformation to martensite could not be obtained at $-78°C$. On the other hand, if the alloy content was too low, the high M_s–M_f temperature range meant that substantial transformation to martensite occurred on cooling to room temperature. However, this problem could be eased to some extent by varying the solution treatment temperature and this facility is illustrated schematically in Figure 5.12(a). In austenitic steels containing about 0.1% C and 17% Cr, complete solution of $M_{23}C_6$ carbides can be obtained at a temperature of 1050°C. In such a condition, both chromium and carbon exercise their full potential in depressing the M_s–M_f range. However, if the solution treatment temperature is reduced, then a proportion of $M_{23}C_6$ carbide is left out of solution and the transformation range is raised due to reduced levels of chromium and carbon in solid solution. Thus the use of low solution treatment temperatures provides a means of accommodating casts of steel in which the alloy content was slightly higher than the optimum but, even so, this approach still necessitated very tight control over the solution treatment temperature.

In addition to the use of refrigeration treatments, controlled transformation steels could also be hardened by aging or *primary tempering* at a temperature of 700°C. This resulted in the precipitation of $M_{23}C_6$ carbide and, as illustrated in Figure 5.12(b), this causes a substantial increase in the M_s-M_f range due to the reduced level of alloying elements in solid solution. Thus on cooling from the tempering treatment, the steel transforms to martensite. In a completely austenitic steel, the precipitation of carbide is essentially restricted to the grain boundaries and this limits the degree to which the transformation range can be raised. However, if delta ferrite is present in the microstructure, the ferrite–austenite interfaces provide further sites for carbide precipitation and transfor-

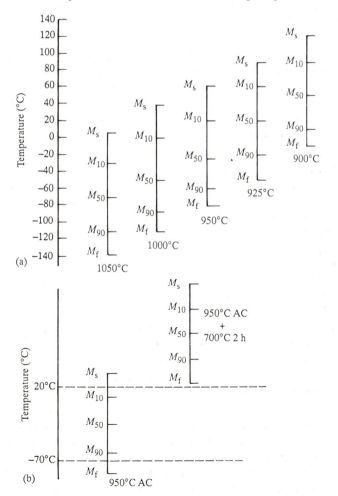

Figure 5.12 *Effect of heat treatment on martensite transformation range: (a) effect of solution treatment; (b) effect of primary tempering (After Irvine et al.[1])*

mation to martensite is produced more readily. Thus where hardening is to be achieved by primary tempering, rather than refrigeration, the steels should contain 5–10% delta ferrite. However, the primary tempering route to transformation results in the formation of lower carbon martensites and lower strength levels compared with those achieved after transformation by refrigeration. This is illustrated in Table 5.6, which shows the tensile properties obtained in a particular grade of steel after the two forms of hardening. This table also shows the low proof stress value and high tensile strength obtained in the solution-treated 1050°C AC condition, which gives an indication of the high rate of work hardening in these meta-stable grades.

As indicated in Table 5.6, these steels are tempered after the hardening treatments and therefore it is desirable to incorporate alloying elements that will improve the tempering resistance or provide a secondary hardening response. Thus many of the commercial grades that were introduced in the 1950s contained about 2% Mo and elements such as copper, cobalt and aluminium were also added in order to promote secondary hardening reactions in the low-carbon martensite. The effect of copper additions on the secondary hardening behaviour of a 17% Cr 4% Ni steel is shown in Figure 5.13 and examples of the compositions of commercial grades of controlled transformation stainless steels are shown in Table 5.7. Although the range of composition shown in this table is quite wide, the steels embody the following features:

Table 5.6   *Properties of a 0.1% C 17% Cr 4% Ni 2% Mo 2% Cu steel*

Heat treatment	TS (N/mm^2)	0.2% PS (N/mm^2)	El % $4\sqrt{A}$
1050°C 1 h AC	1294	213	27.1
1050°C 1 h AC, 700°C 2 h	1035	689	15.6
1050°C 1 h AC, 700°C 2 h, 450°C 6 h	1161	961	19.6
1000°C 1 h AC, −78°C 2 h	1347	874	15.6
1000°C 1 h AC, −78°C 2 h, 450°C 6 h	1340	1154	22.5

Table 5.7   *Compositions of commercial grades of controlled-transformation stainless steels*

Grade	C%	Mn%	Si%	Cr%	Ni%	Mo%	Cu%	Co%	Al%	Ti%
Armco 17-4PH	0.04	0.5	0.3	16.5	3.5		3.5			
AM350	0.1	0.5	0.3	17	4.2	2.75				
Stainless W	0.07	0.5	0.3	17	7				0.2	0.7
Armco 17-7PH	0.07	0.5	0.3	17	7				1.1	
Armco 17-5PH	0.07	0.5	0.3	15	7	2.5			1.2	
FV 520(S)	0.07	1.5	0.5	16	5.5	2	1.5			0.1
SF 80T	0.08	0.2	2.0	17	4	1	1.2	2		

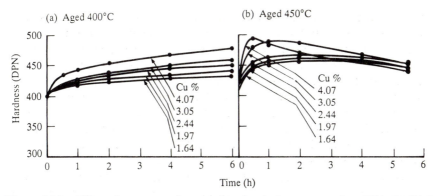

Figure 5.13 *Effect of copper on the secondary hardening response in a 17% Cr 4% Ni steel (initial conditions 950°C AC, − 78°C, 1 h) (After Irvine* et al.[1])

1. A chromium content of 15–17% in order to provide good corrosion resistance.
2. Sufficient austenite-forming elements to produce a mainly austenitic structure at 1050°C.
3. Alloying elements to promote tempering resistance/secondary hardening reactions.
4. An overall alloy content that produces an M_s–M_f transformation range just below room temperature.

These steels were developed primarily for the aerospace industry but were of limited commercial success due to the fact that:

1. The very tight composition ranges were difficult to achieve in commercial production.
2. The meta-stable nature of the steels results in a very high rate of work hardening which limited some aspects of cold formability.

However, FV 520(S) is still in use in the UK and is used primarily in defence applications such as gun carriages and aircraft.

Steel prices

At the time of writing (March 1994), stainless steel producers were facing extremely poor trading conditions and the pricing situation was described as 'dynamic'. Rather than quote prices for this text, in terms of £/tonne, they provided cost ratios for various grades based on a value of *1.0* for Type 304 (18% Cr 9% Ni) cold rolled strip. This information is presented in Figure 5.14 which differentiates between the prices for cold reduced strip and moderately thick plate. Very clearly, the prices reflect the alloy content of the steels, Type 409 (12% Cr) being the cheapest and Type 310 (25% Cr 20% Ni) the most expensive in this particular selection of grades. In turn, Type 316 (18% Cr 12% Ni 2.75% Mo) is more expensive than Type 304 (18% Cr 9% Ni) and a premium is charged for the low carbon (L grades) and Ti-stabilised versions of these grades.

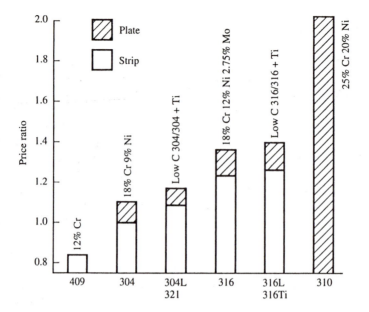

Figure 5.14   *Prices of stainless steel strip and plate relative to a value of 1.0 for Type 304 cold reduced strip at 1 March 1994*

Purely as a guide, a stainless steel producer has indicated that Type 304 cold reduced strip is currently being sold at £1450/tonne which is about 20% lower than the price quoted in the previous edition of this text (January 1991). At that time, the price of plate was lower than that for cold reduced strip but the situation is now reversed. The reasons for this reversal are complex but involve a combination of changes in production methods and competitive pricing between various producers.

The prices of some high-alloy stainless steels, relative to Type 316L, are shown in Table 5.9.

Corrosion resistance

Stainless steels owe their corrosion resistance to the formation of the so-called *passive film*. This consists of a layer, only 20–30 Å thick, of hydrated chromium oxide (Cr_2O_3), which is extremely adherent and resistant to chemical attack. If the passive film is damaged by abrasion or scratching, a healing process or *repassivation* occurs almost immediately. In general, the corrosion and oxidation resistance of stainless steels increase with chromium content and the materials are used in a wide range of aggressive media in the chemical and process plant industries. However, under certain conditions, stainless steels are susceptible to highly localized forms of attack in relatively mild environments, rendering them unsuitable for further service. In this context, the main types of localized attack are *intergranular corrosion, pitting corrosion* and *stress corrosion*. However, it should be stressed that these forms of attack are well researched and thoroughly documented and therefore it is now rare for them to lead to premature or unexpected failure in stainless steel components.

Intergranular corrosion

Given favourable temperature conditions, solute atoms can segregate to the grain boundaries, causing enrichment in a particular element or the precipitation of metal compounds. Under highly oxidizing conditions, these effects can cause the grain boundaries of stainless steels to become very reactive, leading to the highly localized form of attack known as *intergranular corrosion*.

Both austenitic and ferritic stainless steel are susceptible to intergranular corrosion and, in either case, the problem is caused by the segregation of carbon to the grain boundaries and the formation of the chromium-rich, $M_{23}C_6$ carbides. The concentration of chromium in these carbides is very much higher than that in the surrounding matrix and, at one time, it was postulated that this resulted in galvanic corrosion between the noble carbides and the more reactive matrix. However, currently the most widely accepted theory for intergranular corrosion in stainless steels is that involving *chromium depletion*. Thus, in forming chromium-rich carbides at the grain boundaries, chromium is drawn out of solid solution, and in areas adjacent to the boundaries, the chromium content becomes severely depleted compared to the bulk chromium concentration of the steel. Such areas are then said to have become *sensitized* in that they no longer contain sufficient chromium to withstand corrosive attack. Corrosion can then proceed along the grain boundaries and a micrograph illustrating this form of attack in an austenitic stainless steel is shown in Figure 5.15.

Two main laboratory tests are carried out for the evaluation of intergranular corrosion, namely the Huey and Strauss tests. The former was developed to

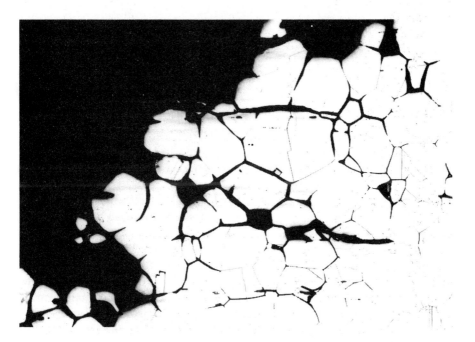

Figure 5.15 *Intergranular corrosion in Type 304 stainless steel*

determine the performance of stainless steels in nitric acid plant and has been standardized as ASTM A262-70 Practice C. It consists of immersing a sample in boiling 65% HNO_3 for five periods, each of 48 hours, using fresh acid solution in each period. The samples are weighed after each period and the weight loss is generally converted to a corrosion rate in terms of mm/year. The geometry of the test piece has to be carefully controlled to avoid excessive 'end grain' effects, which will be described later, and the type of acid used must also be controlled very carefully in order to obtain reproducible results.

The Strauss test was introduced to determine the pickling behaviour of stainless steels and is now standardized as ASTM A262-70 Practice E. This involves exposure to boiling 15.7% $H_2SO_4 + 5.7\%$ $CuSO_4$ solution with the test specimen in contact with metallic copper which increases the severity of attack. The performance of the steel is then judged in terms of the presence or absence of cracks or fissures after bending.

A third test, often designated the *oxalic acid test*, has been standardized as ASTM A262-70 Practice A. In this test, a polished sample of austenitic stainless steel is etched electrolytically in a 10% solution of oxalic acid at room temperature. In the absence of chromium carbide precipitates, the grain boundaries of the etched sample exhibit a *step* between adjacent grains compared with a *ditch* in material which has experienced intergranular attack. The test also recognizes a dual, intermediate condition in which some inter-granular corrosion has occurred but no single grain is surrounded completely by ditches.

At temperatures below about 850°C, the solubility of carbon in an austenitic stainless steel falls below 0.03% and exposure in the temperature range 450–800°C can result in the precipitation of $M_{23}C_6$ carbide. This can occur during:

1. Slow cooling through the sensitization temperature range following solution treatment at 1050°C or after welding.
2. Stress relieving after welding or service exposure in the critical temperature range.

Slow cooling from welding and subsequent intergranular attack gave rise to the term *weld decay* before the mechanism was fully understood. The weld decay area is generally a band of material, some distance from the weld, which has been exposed to the temperature range favouring the precipitation of chromium carbide.

From the foregoing remarks, it will be obvious that a reduction in the carbon content will reduce the susceptibility to intergranular corrosion but, at one time, major expense was incurred in reducing the carbon content of an austenitic stainless steel to below 0.06%. However, as indicated earlier, the problem was controlled initially by adding elements such as titanium or niobium, which are stronger carbide formers than chromium, and therefore TiC or NbC are formed rather than the damaging chromium-rich $M_{23}C_6$. The addition of these elements is said to *stabilize* stainless steels against intergranular attack and gives rise to the standard grades – Type 321 (Ti-stabilized) and Type 347 (Nb-stabilized). These elements also form nitrides and the additions made to stainless steels are slightly in excess of those required by stoichiometry for complete precipitation of carbon and nitrogen, namely, $Ti = 5 \times (C + N)\%$ and $Nb = 10 \times (C + N)\%$.

Although the problem of intergranular corrosion is controlled by the addition of stabilizing elements, Types 321 and 347 are not completely immune to this form of corrosion since they are susceptible to *knife-line attack*. During welding, the heat-affected zone is raised to temperatures above 1150°C and this can result in the partial dissolution of TiC and NbC. Carbon is therefore taken into solution in a narrow region adjacent to the weld and can be available for the formation of chromium carbide on cooling through the sensitization range of 450–800°C. The susceptible region may only be a few grains wide but can give rise to a thin line of intergranular attack. Hence the term *knife-line attack*.

The problem of intergranular corrosion in austenitic stainless steels can also be overcome by reducing the carbon content to low levels. Although steels containing 0.03% C max. (*L grades*) have been available since the 1950s, their use was restricted very severely in the early days due to high production costs. This involved the use of low-carbon ferro-chrome, which was expensive. However, the introduction of AOD refining has cheapened the production of the *L-grades* very considerably and these steels are now used extensively in applications formerly satisfied by the stabilized grades. However, the presence of TiC and NbC particles gives rise to some dispersion strengthening and Types 321 and 347 are still used in applications where advantage can be taken of their higher strength.

According to Sedriks,[8] molybdenum has an adverse effect on intergranular corrosion as assessed in the Huey test and a beneficial effect in the Strauss test. It is suggested that the adverse effect in the former may be concerned with the fact that nitric acid attacks areas other than chromium-depleted regions, e.g. areas associated with solute segregation or the early formation of sigma phase. At one time, small additions of boron were made to austenitic stainless steels in order to improve hot workability and the high-temperature creep properties. However, such additions are very deleterious to the performance in the Huey test. It has been suggested that boron additions lead to the formation of M_2B and $M_{23}(CB)_6$ precipitates at the grain boundaries, both of which give rise to chromium-depletion effects. Performance in the Huey test is also improved by reducing the phosphorus content of austenitic stainless steels to low levels, i.e. less than 0.01% P.

In addition to segregation and precipitation effects at grain boundaries, the intergranular corrosion behaviour of austenitic stainless steels can also be influenced very markedly by the presence of elongated particles or clusters of second phases. These can take the form of sulphides or other plastic inclusions but adverse effects can also be induced in niobium-stabilized Type 347 steel due to the presence of stringers of coarse niobium carbonitrides. The mechanism is shown schematically in Figure 5.16. In the Huey test, outcropping inclusions appear to dissolve quickly, creating long narrow passages from the end faces into the body of the samples. These passages do not provide ready access for the ingress of new acid and it is thought that this gives rise to the formation and concentration of hexavalent Cr^{6+} ions. This leads to particularly aggressive corrosion conditions within the passages and intergranular attack proceeds very

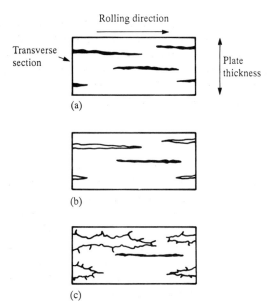

Figure 5.16   *Schematic illustration of end grain attack from outcropping inclusions: (a) elongated inclusions outcropping on transverse sections; (b) outcropping inclusions dissolved in nitric acid; (c) heavy intergranular attack from crevices*

quickly along the adjacent grain boundaries. The effect is known as *end grain attack* and can lead to weight losses far in excess of that experienced in cleaner steels or those containing shorter inclusions. In chemical plant, end grain corrosion can be experienced in forgings or in tube-to-tube welds where end faces are exposed due to ovality effects or differences in tube diameter. In such cases, the exposed end grain faces should be covered by capping welds.

Because carbon has a very low solid solubility in ferritic stainless steels, it is extremely difficult to prevent the formation of chromium carbides in these grades. Thus when carbon is taken into solution by heating to temperatures above 925°C, carbide precipitation may occur even when the material is water quenched from the solution treatment temperature. However, intergranular corrosion will only proceed in sensitized material in aggressive, oxidizing media and, ordinarily, ferritic stainless grades such as Types 430 and 434 are not exposed to such environments. Even so, care should be taken to minimize the risk of intergranular corrosion.

Ferritic stainless steels sensitize more rapidly and at lower temperatures than their austenitic counterparts, the fastest reaction occurring at a temperature of about 600°C. However, by holding at a temperature of about 800°C, or by cooling slowly through the temperature range 700–900°C, the risk of intergranular attack can be eliminated. Following the initial, damaging precipitation of chromium carbides, these treatments allow sufficient time for chromium to diffuse into the depleted zones and thereby eliminate the sensitization effects.

The addition of titanium also reduces the risk of intergranular attack in ferritic stainless steels. However, whereas the reduction of carbon to below

0.03% is effective in preventing sensitization in austenitic grades (L grades), significantly lower levels are required in ferritic stainless steels. Nitrogen is also damaging, causing chromium depletion via the formation of chromium nitride. Therefore, the total (carbon + nitrogen) must be reduced to below 0.01% in order to prevent sensitization in a 17% Cr ferritic steel. The development of *low interstitial ferritic steels* will be discussed later.

Pitting corrosion

As its name suggests, pitting is a highly localized form of corrosion which, in its initial form, results in the formation of shallow holes or pits in the surface of the component. However, the pits can propagate at a fast rate, resulting in *pin-holing* or complete perforation in the wall of the component. Therefore pitting can be completely destructive in terms of further useful life when only a very small amount of metal has been attacked by corrosion. In stainless steels, pitting corrosion generally takes place in the presence of chloride ions and it is widely held that the initiation stage is associated with attack on non-metallic inclusions. However, other microstructural features may also play a part. The formation of a pit in an austenitic stainless steel is shown in Figure 5.17.

Fontana and Greene[9] have stated that pits usually grow in the direction of gravity, i.e. downwards from horizontal surfaces, and only rarely do they proceed in an upward direction. These authors have formulated a model for pitting in terms of an autocatalytic process for stainless steels in aerated sodium chloride solution. The sequence of events is as follows:

1. Anodic dissolution takes place at the bottom of the pit:

$$M \rightarrow M^+ + e^-$$

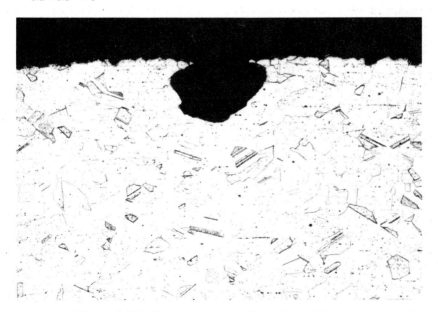

Figure 5.17 *Pitting corrosion in Type 304 stainless steel*

2. A cathodic reaction takes place on adjacent surface:

$$O_2 + 2H_2O + 4e^- \rightarrow 4OH^-$$

3. The build up of M^+ ions within the pit causes the Cl^- ions to migrate to the pit in order to preserve electrical neutrality.

4. The soluble metal chloride hydrolyses to form hydroxide and free acid:

$$M^+ Cl^- + H_2O \rightarrow MOH + H^+Cl^-$$

Thus acid is produced by the reaction which decreases the pH to a low level whereas the bulk solution remains neutral.

Traditionally, the susceptibility of stainless steels to pitting corrosion was evaluated by immersion tests in acidified ferric chloride solutions (ASTM Practice G48-76 and Practice G46-76). Such tests involve the measurement of pit density, size and depth. However, laboratory tests on pitting behaviour are now more generally based on electrochemical techniques.

The resistance to pitting increases with chromium content but major benefit is obtained from the addition of molybdenum in stainless steels, e.g. Type 316 (18% Cr, 12% Ni, 2.5% Mo). The addition of nitrogen is also beneficial and the combined effects of nitrogen plus molybdenum will be discussed later in this chapter. The potential pitting resistance of stainless steels is often expressed in terms of a pitting index:

$$\text{Pitting index} = Cr\% + 3.3Mo\% + 16N\%$$

In terms of microstructure, MnS inclusions are important sites for pit initiation but other features such as delta ferrite, alpha prime and sigma phase can also promote pitting corrosion.

Stress corrosion cracking

Stress corrosion cracking (SCC) is a form of failure induced by the conjoint action of tensile stresses and particular types of corrosive environments. The stresses can be either applied or residual and cracking takes place in a direction normal to the tensile stresses, often at stress levels below the yield strength of the material. A micrograph illustrating SCC in an austenitic stainless steel is shown in Figure 5.18. Cracking can take place in either a transgranular or intergranular manner and can proceed to the point where the remaining material can no longer support the applied stress and fracture then takes place.

SCC occurs in chloride, caustic or oxygen-rich solutions but the majority of failures in austenitic stainless steel take place in chloride-bearing environments. Chloride ions may be present in the process stream but can also be introduced accidentally through the incomplete removal of sterilizing agents, such as hypochlorite solutions.

It is now generally acknowledged that SCC is initiated by the formation of a pit or crevice at anodic sites on the surface, e.g. outcropping inclusions which disrupt the passive film. The reactions controlling pitting corrosion were

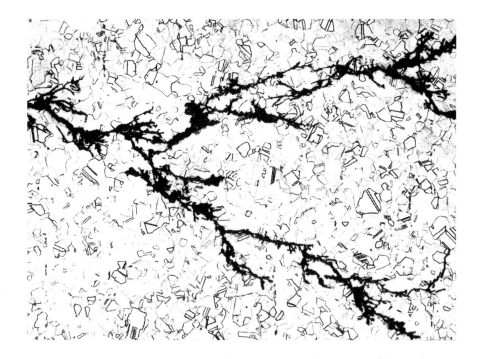

Figure 5.18 *Stress corrosion cracking in Type 304 stainless steel*

described in the previous section. A number of mechanisms has been proposed
for crack propagation but the surviving theories are based on anodic dissolution
and hydrogen embrittlement mechanisms. In the former, cracking propagates
by local anodic dissolution at the crack tip, passivation at that point being
prevented by plastic deformation. Alternatively, the principal role of plastic
deformation is to accelerate the dissolution process. In the hydrogen embrittlement theory, hydrogen is generated due to the cathodic reaction ($H^+ + e^- \rightarrow H$)
and migrates to the crack tip where it is absorbed within the metal lattice,
causing mechanical rupture. This extends the crack and further anodic dissolution then occurs on the newly exposed surfaces.

Although laboratory tests are carried out in a variety of salt solutions, that
involving boiling 42% magnesium chloride solution is probably the most
popular. The test specimens may be U-bend specimens which are clamped so as
to provide a large tensile stress on the outside of the bend or else tensile
specimens are employed, loaded to perhaps 0.8 times the 0.2% PS of the
material.

Major research effort has been devoted to the effects of composition on the stress corrosion behaviour of stainless steels. One of the most important elements is nickel and the celebrated Copson curve[10] is shown in Figure 5.19. This relates to stainless steels containing 18–20% Cr in a magnesium chloride solution boiling at 154°C. This figure indicates a minimum resistance to SCC at nickel contents in the range 8–10%, i.e. typical of those in standard austenitic steels such as Types 302 and 304. A reduction in nickel from these levels leads to an improvement in the SCC behaviour but of course this is associated with the progressive replacement of austenite by delta ferrite in the microstructure. Alternatively, the resistance to SCC can be improved by increasing the nickel content to levels of 40% and above. The beneficial effect of nickel at levels above 8–10% has been attributed to an increase in the *stacking fault energy* of the materials but nickel must also exert an ennobling effect which inhibits anodic dissolution.

Silicon has a beneficial effect on the SCC behaviour in magnesium chloride solutions but has virtually no effect in high-temperature solutions of sodium chloride.

Although clearly beneficial in the formation of passive films and in suppressing pitting corrosion, both chromium and molybdenum appear to have a variable effect on the SCC behaviour of stainless steels. These effects may well be linked to austenite stability and a reduction in stacking fault energy which promotes crack propagation.

Temperature is a very important parameter in the SCC behaviour of stainless steels and is rarely observed at temperatures below about 60°C.

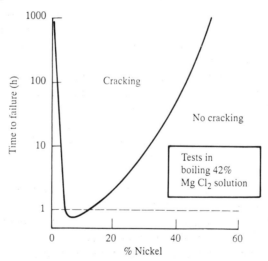

Figure 5.19 *Effect of nickel on the stress corrosion behaviour of austenitic stainless steels (After Copson[10])*

High-alloy stainless steels

As indicated at the beginning of this chapter, most of the popular grades of stainless steel were formulated in the 1920s and 1930s and have undergone little change to the present day. However, in the intervening period, a large amount of research effort has been devoted to the development of new grades but relatively few of these compositions have enjoyed major commercial exploitation.

High-alloy austenitic/duplex grades

From the basic Type 304 (18% Cr, 9% Ni) composition, a number of developments have taken place which were designed to improve a particular property. These include:

1. The addition of up to 0.5% N to improve the strength (solid solution strengthening) or to compensate for reductions in the nickel content (e.g. Types 201 and 202).
2. The addition of elements such as aluminium, titanium or phosphorus to produce precipitation-strengthening reactions.
3. The inclusion of high levels of nickel and molybdenum and moderate levels of copper and nitrogen to improve the resistance to stress corrosion or reducing acids.

The properties of steels in the first category were discussed briefly on page 230. In contrast to the situation in low-carbon strip and structural grades, very little use has been made of *Hiproof*-type compositions which contain about 0.2% N and provide 0.2% PS values of about 330 N/mm^2 (cf. 250 N/mm^2 in standard grades). Having overcome the welding problems that were encountered when these steels were first introduced, it is surprising that they are not yet being used in large quantities. Whereas the 200 series have enjoyed commercial success in the United States, they have always been regarded as inferior substitutes to the traditional 300 steels in the UK and, as stated earlier in the text, they also present problems in production.

The metallurgy of precipitation-strengthened austenitic stainless steels was the subject of detailed research work in the 1950s and 1960s.[11] Depending upon the precipitation system and heat treatment, such steels can develop 0.2% PS values up to 750 N/mm^2, i.e. three times that produced in Type 304 in the solution-treated condition. However, such grades were produced in very limited quantities, probably for the following reasons:

1. The lack of facilities and expense involved in carrying out the aging treatments – typically 700–750°C for 24 hours.
2. The reduction in corrosion resistance produced by the precipitation of carbides, phosphides or intermetallic compounds.
3. The loss of strength in the heat-affected zones of welds.
4. The reluctance of plant designers to consider the very large reductions in plate thickness etc. that could be provided by these high-strength steels.

As illustrated in Table 5.8, a large number of standard and proprietary grades have been developed with the objective of increasing the resistance to reducing acids, pitting and crevice corrosion. Many of the steels can be regarded as simple extensions of Type 316 (18% Cr, 12% Ni, 2.5% Mo) with higher molybdenum contents and with increased levels of nickel to preserve a fully austenitic structure. Some of these steels also contain up to 0.25% N which increases the strength and augments the chromium and molybdenum contents in improving the pitting resistance, according to the following pitting index formula:

Pitting index $= Cr + 3.3\%\ Mo + 16\%\ N$

Copper additions of up to 2% feature in some of the compositions which improve the resistance to sulphuric acid.

Many of the compositions listed in Table 5.8 are used in flue gas desulphurization plant where acid condensates, contaminated with chlorides, produce severe corrosive conditions. Steels such as 94L are also used in the food industry and also for heat exchangers, particularly where salt water is involved. This grade contains 25% Ni and will therefore provide better resistance to stress corrosion than traditional grades such as Type 316.

The last two compositions in Table 5.8 differ from the rest in that the high chromium and molybdenum contents are not complemented with a large addition of nickel and this results in a duplex, austenite plus ferrite structure. Such steels have good resistance to both chloride and sulphide stress corrosion and the duplex structure also provides high proof stress values, e.g. 480 N/mm^2.

Table 5.8 *Compositions of high-alloy austenitic and duplex stainless steels*

Alloy designation	Chemical composition (wt %)						
	C	*Cr*	*Mo*	*Ni*	*Cu*	*N*	*Others*
316L	0.03	17	2.25	12	–	–	–
316LM	0.03	17	2.75	12	–	–	–
317L	0.03	18	.5	15	–	–	–
317LN	0.03	18	4.5	13	–	0.15	
317LM	0.03	18	4.5	15	–	–	–
317LMN	0.04	18	6	14	–	0.15	2.1Nb + Ta,
94L	0.02	20	3	25		–	2.5Co, 1W
254SMo	0.02	20	6.5	18	2	0.12	5Co, 1.5W
Incoloy 825	0.03	22		42		–	3.65Nb + Ta
Hastelloy G	0.05	22	7	44	1.9	–	3W
			9		–		1.25Co, 4W
Hastelloy G-3	0.015	22	13	41	–	–	–
Inconel 625	0.05	22	16	61	–	–	0.75W
Hastelloy C-22	0.015	22	3		–	–	–
Hastelloy C-276	0.02	25	3.5		0.75	–	–
22/5	0.03					0.15	
Zeron 100	0.03					0.25	

After Lane and Needham.[12]

This combination of properties is used to advantage in the petroleum industry, particularly for the handling of crude oil which is contaminated with hydrogen sulphide. Typical applications include transfer pipelines, downhole tubing and liners and also topside process equipment, such as heat exchangers and seawater transport piping.[13]

British Steel[13] has presented the information in Table 5.9 on the prices of some of the high-alloy grades and nickel superalloys relative to Type 316L.

Table 5.9

Alloy types	Price factor
316L	1
22/5	1.1
317L	1.2
317LM	1.3
94L	2
High-nickel alloys	2.3–3
High-molybdenum alloys ($\geqslant 6\%$ Mo)	2.7–3.8
Nickel-base alloys	3.3–9
Titanium	7.6–8.1

High-alloy ferritic grades

As indicated earlier, standard austenitic stainless steels such as Type 304 (18% Cr, 9% Ni) are highly susceptible to chloride stress corrosion and the nickel content must be raised to about 40% in order to provide immunity to this form of attack. Such materials are expensive and the addition of nickel alone provides little benefit in terms of resistance to pitting and crevice corrosion. On the other hand, standard ferritic grades such as Type 430 (17% Cr) and 446 (25% Cr) are resistant to stress corrosion but are prone to intergranular corrosion and have poor toughness, even in moderate plate thicknesses. However, as early as 1951,[14] it was shown that these deficiencies could be overcome by reducing the interstitial content of ferritic stainless steels to low levels. However, for many years, this information remained little more than laboratory-based data until the development of steelmaking processes that enabled the materials to become commercially viable. In particular, the Airco-Temescal Division of Airco Reduction Co. Inc. in the United States introduced a vacuum melting process with electron beam stirring which could achieve total carbon and nitrogen contents of the order of 0.01%.[15] In addition, an LD converter coupled with a Standard-Messo decarburizer was developed in Germany to achieve total interstitial contents of about 0.025%.[16] However, the argon–oxygen decarburization (AOD) process has also been used for the production of these grades. Thus in the mid-1960s major commercial interest was stimulated in low interstitial ferritic stainless steels which offered the potential of cheaper alternatives to austenitic grades, particularly in relation to resistance to stress corrosion.

The influence of carbon and nitrogen on the impact properties of an 18% Cr 2% Mo base steel was investigated by Hooper *et al.*[17] and the data are shown in Figure 5.20. From this figure, it is apparent that a reduction in carbon content to 0.01% produces a marked improvement in the impact transition behaviour, whereas a change in nitrogen content has little effect. However, the grain size of these steels was about ASTM 0 and even at low interstitial levels, the impact transition temperature is still above room temperature. Therefore, in order to produce attractive impact properties, the grain size must be reduced, i.e. to a level below ASTM 8.

In addition to a reduction in carbon and nitrogen contents, the new range of ferritic steels contained higher levels of chromium and additions of molybdenum and nickel. Streicher[18] has reviewed the development of these grades and the types of composition that have been produced in commercial quantities are shown in Table 5.10. Although nitrogen has a relatively small effect on the impact properties, its low solubility in ferritic grades means that chromium nitride forms very readily and contributes to the susceptibility to intergranular attack. Thus many of the high-alloy ferritic grades must be stabilized with titanium or niobium, even when the carbon and nitrogen has been reduced to very low levels.

As illustrated in Table 5.11, most of these grades are resistant to stress corrosion cracking in MgCl₂ and NaCl solutions and they also display good resistance to pitting and crevice corrosion. The inclusion of nickel in steels such as 28% Cr 2% Mo 4% Ni–Nb and 29% Cr 4% Mo 2% Ni is beneficial in improving the corrosion resistance in reducing acids. It is also claimed[18] that nickel improves the toughness of ferritic stainless steels.

Low interstitial ferritic stainless steels have enjoyed some commercial success in Sweden and the United States but failed to make much impression in the UK.

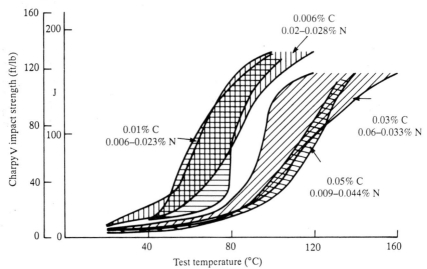

Figure 5.20 *Effect of carbon and nitrogen on the impact properties of 18% Cr 2% Mo steel (After Hooper* et al.[17])

Table 5.10 *Commercial grades of low interstitial ferritic stainless steels (After Streicher[18])*

	Alloy	Limits for carbon, nitrogen and stabilizers (%)	Melting and refining processes
I	Fe-18 Cr-2 Mo-Ti	C – 0.0250 max. N – 0.0250 max. C + N < 0.030 desirable Ti + Nb = 0.20 + 4 (C + N) min. = 0.80 max.	Argon–oxygen decarburization or vacuum–oxygen decarburization
II	Fe-26 Cr-1 Mo	C – 0.0050 max. N – 0.0150 max. Nb = 13 to 29 (N)	Electron beam hearth refining or vacuum induction melting
II-A	Fe-26 Cr-1 Mo-Ti	C – 0.0400 max. N – 0.0400 max. 0.2 to 1.0 Ti C + N = 0.050 typical	Argon–oxygen decarburization
III	Fe-28 Cr-2 Mo	C + N ⩽ 0.0100 Desirable C + N ⩽ 0.050	Vacuum melting followed by arc remelting
III-A	Fe-28 Cr-2 Mo-4 Ni – Nb	C – 0.0150 max. N – 0.0350 max. C + N ⩽ 0.0400 with Nb ⩾ 12 (C + N) + 0.2	Vacuum–oxygen decarburization Argon–oxygen decarburization
IV	Fe-29 Cr-4 Mo	C – 0.0100 max. N – 0.0200 max. C + N ⩽ 0.0250 max.	Vacuum induction melting or electron beam refining
IV-A	Fe-29 Cr-4 Mo-2 Ni	Same	Vacuum induction melting

Table 5.11 *The properties of low interstitial ferritic stainless steels (After Streicher[18])*

Property	Alloy					
	18Cr-2Mo-Ti	26Cr-1Mo (High purity)	w26Cr-1Mo-Ti	28Cr-2Mo-4Ni-Nb	29Cr-4Mo	29Cr-4Mo-2Ni
Stress Corrosion Cracking						
$MgCl_2$ (155 C, 310 F)	R*‡	R*	R*95	F	R*	R
NaCl (103 C, 217 F)	R	R	R	R	R	R
Pitting and Crevice Corrosion						
$KMnO_4$-NaCl						
Room Temperature	F	R	R	R	R	R
50 C (120 F)	—	R	R	R	R	R
90 C (195 F)	—	F	F	F	R	R
Ferric Chloride						
Room Temperature	F	F	F	R	R	R
50 C (120 F)	—	—	—	F	R	R
Corrosion in Boiling Acids[1]						
Nitric–65%	F†	R	F†	R	R	R
Formic–45%	F	R	R	R	R	R
Oxalic–10%	F	F	F	R	R	R
Sulfuric–10%	F	F	F	R	F	R
Hydrochloric–1%	F	F	F	R	F	R
Transition Temperature[2]	+25 to +75 C (+75 to +165 F)	−62 C (−80 F)	+40 C (+105 F)	−5 C (+25 F)	+16 C (+60 F)	−7 C (+20 F)
Refining Processes	AOD or VOD	EB or VIM	AOD	VOD or AOD	VIM or EB	VIM

[1] R indicates passive or self-repassivating with rates <0.2 mm/yr 0.008 in./yr

[2] Full size Charpy V-notch specimens

* Copper and nickel residuals must be kept low in these alloys to resist cracking in this solution

† Not recommended for oxidizing solutions

‡ Data from Climax Molybdenum Co. publication, "18Cr-2Mo Ferritic Stainless Steel"

R = Resistant
F = Fails by type of corrosion shown
AOD = Argon–Oxygen Decarburization
VOD = Vacuum–Oxygen Decarburization
VIM = Vacuum Induction Melting
EB = Electron Beam Refining

However, the author recalls one trial, involving an 18% Cr 2% Mo steel, for the manufacture of a brewery vessel. This trial proved to be unsuccessful due to fabrication problems and poor impact properties in the heat-affected zones of welds.

Welding of stainless steels

A very high proportion of stainless steels is welded in the fabrication of pressure vessels, storage tanks, chemical plant and domestic appliances. In each case, the welds are required to be of high integrity and to provide corrosion resistance or mechanical properties similar to that of parent material. Therefore good weldability is a particularly attractive feature of stainless steels and this contributes very significantly to the wide usage and versatility of these materials. The welding of stainless steels has been reviewed extensively by Castro and de Cadenet[19] and the topic has been updated more recently by Gooch.[20]

Martensitic stainless steels

Because of their high alloy content, 12% Cr steels and other martensitic stainless grades exhibit a high level of hardenability and are capable of transforming to martensite in large section sizes. Therefore these steels are susceptible to cold cracking in the weld metal and HAZ in a similar manner to low-alloy martensitic grades. As with their low-alloy counterparts, the problem is exacerbated by the presence of hydrogen, and the risk of cracking increases as the carbon content and hardness of the steels are increased. For this reason, the welding of 12% Cr martensitic grades is generally restricted to compositions containing a maximum of about 0.25% C. However, the problem of hydrogen cracking is well understood and can be readily overcome in martensitic stainless steels provided the normal precautions are taken. Thus in gas-shielded processes, pure argon should be used rather than argon–hydrogen mixtures, and in manual metal arc welding, low-hydrogen basic electrodes should be employed. The thorough baking of electrodes is also recommended in order to remove the last traces of moisture.

Like other highly hardenable grades, 12% Cr steels are generally preheated to reduce the risk of cracking. Preheating temperatures of up to 250°C are employed which ensure that the weld metal and HAZ cool slowly through the M_s–M_f temperature range, thereby facilitating stress relaxation and reducing thermal stresses. On the other hand, martensitic stainless steel containing less than 0.1% C and with nickel additions of up to 4% are often welded without preheat. However, it is still imperative that low levels of hydrogen are produced in the weld metal with matching fillers so as to avoid weld metal cracking. The use of Cr–Ni austenitic filler metal can also remove the need for preheating, although the strength of the joint will be lower than that achieved with martensitic weld metals.

After welding, 12% Cr steels are subjected to heat treatment in order to provide an adequate balance between strength and toughness in the weld zone.

Because these steels have a low M_s–M_f temperature range, austenite may well be present immediately after the completion of welding and it is essential that the assembly is cooled to room temperature in order to complete the transformation before applying post-weld heat treatment. If adequate cooling is not carried out, austenite will be present in the microstructure during the subsequent tempering treatment and this may well transform to untempered martensite on cooling to room temperature. Post-weld heat treatments are generally carried out in the range 600–750°C.

Martensitic stainless steels can be welded using all the conventional fusion processes, providing precautions are taken to minimize the level of hydrogen in the weld.

Austenitic stainless steels

Austenitic stainless steels are readily welded, given their high level of toughness and freedom from transformation to martensite. Therefore they are not prone to the cold-cracking problems encountered in martensitic stainless steels and require neither preheating nor post-weld heat treatment. Because of these characteristics, austenitic fillers are often used for joining dissimilar steels or, as mentioned earlier, in the welding of brittle martensitic stainless grades. However, precautions must be taken to avoid other problems in austenitic stainless steels, namely *solidification cracking* and *sensitization*.

Solidification cracking takes place in the weld metal as it is about to solidify. The problem is due to the generation of high contraction stresses in austenitic stainless steels because of the high thermal expansion characteristics of these materials. These stresses pull the solidified crystals apart when they are still surrounded by thin films of liquid metal, giving rise to interdendritic cracking. However, it was discovered at an early stage that welds with a fully austenitic structure were particularly susceptible to solidification cracking and that the problem could be overcome with the introduction of a small amount of delta ferrite. The amount of ferrite required to eliminate cracking depends upon the degree of restraint imposed upon the joint and also on the composition of the steel. However, a level of about 5% ferrite is generally adequate. Therefore in autogenous welds, the compositions of commercial grades such as Types 304, 316, 321 and 347 are balanced such that they are free of delta ferrite in the solution-treated condition but generate about 5% ferrite in the weld metal due to the thermal excursion into the liquid state. Similarly, in the joining of thick sections of austenitic stainless steels, filler metals of the correct composition are employed which develop the required microstructure.

The prediction of the microstructure in the weld metal of stainless steels can be carried out using diagrams prepared by Schaeffler[21] and De Long.[22] The latter was developed later and has the advantage that it takes account of the nitrogen content of the steel. As indicated earlier, nitrogen is a powerful austenite-forming element and can therefore have a significant effect on the microstructure. The De Long diagram is shown in Figure 5.21 and the chromium and nickel equivalents are calculated in the following manner:

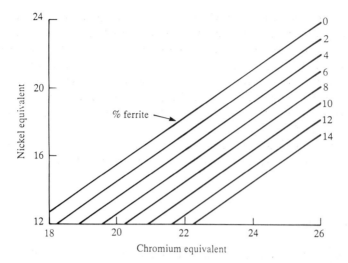

Figure 5.21   *De Long diagram for prediction of microstructure in stainless steel welds (After De Long[22])*

Cr equivalent $= \%Cr + Mo + 1.5Si + 0.5Nb$
Ni equivalent $= \%Ni + 30C + 30N + 0.5Mn$

Solidification cracking is promoted by the presence of certain elements which segregate to the remaining liquid during the solidification process, producing interdendritic films of low melting point. Elements such as nickel, silicon, sulphur and phosphorus increase the susceptibility to cracking whereas chromium, nitrogen and manganese reduce the cracking tendency. With this knowledge, filler metals can be designed which are completely austenitic and yet resistant to solidification cracking. Such fillers are low in sulphur, phosphorus and silicon but often contain 7–10% Mn. One particular use of these *zero ferrite* electrodes is in the welding of grades such as Type 310 (25% Cr 20% Ni) in applications calling for low magnetic permeability, which precludes the introduction of delta ferrite.

The problem of chromium carbide precipitation at grain boundaries and sensitization to intergranular attack was discussed earlier in the section dealing with the corrosion behaviour of stainless steels. However, this problem can now be avoided with the use of stabilized grades such as Types 321 and 347 or with the L grades where the carbon is restricted to 0.03% max.

Ferritic stainless steels

Again, the absence of transformation in ferritic grades eliminates the potential for cold cracking and, due to their moderate thermal expansion characteristics, these steels are generally free from solidification cracking. On the other hand, ferritic stainless steels are prone to grain growth and this leads to low levels of toughness in the weld, particularly in thicker sections. However, sound autogenous welds are produced in Type 430 (17% Cr) steel in thin gauges in applications such as domestic sinks, washing machines and dish washers. Where thicknesses of several millimetres are to be welded, it is preferable to use an austenitic filler such as Type 316 which produces weld metal of high toughness.

Ferritic stainless steels are susceptible to hydrogen cracking, and in gas shielded processes, pure argon rather than argon–hydrogen mixtures should be used.

Variable weld penetration

With the introduction of automatic welding processes, such as orbital TIG for tube joining, major problems were encountered with stainless steels due to variable weld penetration. Thus marked differences in the depth of penetration were observed when steels of nominally the same composition were welded under identical conditions. Alternatively, there was a tendency for the weld to deflect to one side of the joint when steels from different casts were welded together. For this reason, the problem is also known as *cast-to-cast* variability. Although problems of this kind had been observed previously, and also in materials other than stainless steels, they could be accommodated in manual welding by adjustments to power or welding speed. Whereas factors such as arc–metal interactions may be contributory, it is now generally agreed that variable penetration is due to composition-induced, surface energy effects.

Llewellyn *et al.*[23] investigated the problem in commercial tube and plate samples of Type 304L and found a reasonable correlation between current to penetrate and sulphur content. This effect is shown in Figure 5.22 and regression analysis on the pooled data for tube and plate materials yielded the following relationship:

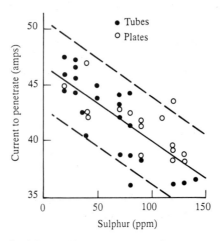

Figure 5.22   *Effect of sulphur on the current required to give penetration of a 2-mm-thick sheet or tube wall with a 5 second arc duration (After Llewellyn et al.[23])*

Current to penetrate (amps) $= 47 - 686$ (%S)

which explained 60% of the variance. No other factors were statistically significant at the 95% confidence level. When similar work was carried out on laboratory casts of Type 304L, sulphur had no effect. However, these materials had oxygen contents of 190–330 ppm compared to 20–70 ppm in the commercial steels. Following the work of Heiple and Roper,[24] it was anticipated that both sulphur and oxygen would be beneficial in promoting good weldability. However, the lack of a sulphur effect in the laboratory casts of Type 304L may well have been due to the overriding effect of oxygen in these materials.

Changes in surface energy across the temperature gradients of the weld pool surface give rise to *Marangoni convection* which has a marked effect on the direction and velocity of fluid flow. Thus direction of fluid flow can be changed by surface active impurities which reverse the negative temperature dependence of surface energy for pure materials to a positive temperature dependence when impurities are present. The effects are illustrated schematically in Figure 5.23(a). The temperature distribution in the weld pool is similar in both 'good' and 'bad' materials (Figure 5.23(b)) but the different relationships between surface tension and temperature in the two materials cause a marked difference in fluid flow (Figure 5.23(c)). Thus in high-purity steels, the surface liquid flow is from the centre to the outside, giving an upward axial flow and poor penetration (Figure 5.23(d)). On the other hand, in steels containing surface active impurities, the flow is towards the centre of the liquid pool, giving downward axial flow and good penetration.

Leinonen[25] has also demonstrated the beneficial effect of sulphur on the TIG weldability of Type 304 steels. He showed that the maximum welding speed that could achieve through-penetration in a steel containing 0.13% S was 74% higher than that for a 0.003% S steel, when using 100% argon shielding gas. The addition of hydrogen and helium to the shielding gas enabled the welding speed to be increased substantially but, in both cases, the differential in weldability between the high- and low-sulphur steels was still preserved.

In order to achieve good weld penetration, fabricators may well specify a minimum of 0.01% S but high levels of sulphur impair the corrosion resistance of stainless steels. Therefore the sulphur content needs to be controlled to a narrow range to ensure good weldability on the one hand and the avoidance of pitting corrosion on the other.

Cold working of stainless steels

The cold-working characteristics of stainless steels are important in relation to:

1. The conversion of hot band to cold-reduced gauges.
2. The production of components by cold forming.

Austenitic stainless steels are characterized by high rates of work hardening which limit the amount of cold deformation that can be undertaken before an

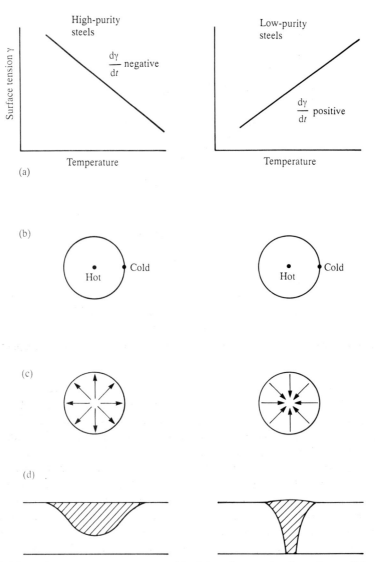

Figure 5.23 *Marangoni convection in welds: (a) surface energy dependence; (b) temperature distribution in welds; (c) fluid flow on surface; (d) bad and good weld penetration*

annealing treatment is required. However, the work-hardening behaviour is influenced markedly by chemical composition and can often dictate the grade of steel that is used for a particular application.

Role of alloying elements

The austenite to martensite transformation in a material can be discussed in terms of thermodynamics and this aspect has been covered in detail by Kaufman and Cohen.[26] In Figure 5.24, F^γ and $F^{\alpha'}$ represent the chemical free

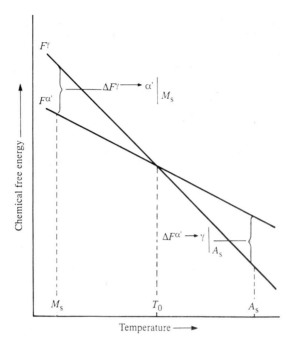

Figure 5.24   *Thermodynamics of the martensite reaction (After Kaufman and Cohen[26])*

energies of austenite and martensite which vary with temperature in the manner shown. At temperature T_0, the two phases have equal free energies, but above T_0, austenite is thermodynamically stable relative to martensite, and below T_0, martensite is the more stable phase. On cooling from above T_0, the austenite to martensite reaction might be expected to occur at T_0 but, in fact, this reaction does not occur until the temperature is lowered to M_s, which in iron-base alloys may be of the order of 200°C below T_0. At M_s, the austenite has derived sufficient energy from undercooling to overcome the barriers of nucleation and the interfacial strains that inhibit the growth of the bcc nucleus in its foreign fcc environment. On heating from a low temperature, the reverse reaction at A_s may also require superheat and A_s is generally as far above T_0 as M_s is below. Alloying elements depress the transformation from austenite to martensite and in very highly alloyed stainless grades, the theoretical T_0 temperature will be below the absolute temperature. However, in the leaner alloy grades, T_0 is above and M_s is below room temperature.

When stress is applied to an austenitic stainless steel during cold working, the mechanical energy interacts with the thermodynamics of the martensite reaction, as demonstrated by Patel and Cohen.[27] Thus in Figure 5.25, F^γ and $F^{\alpha'}$ represent the conditions in the unstressed system and F^γ and $F^{\alpha'}$ are the relative conditions in the stressed system, where F^γ and F^{γ} are displaced by an amount $\Delta G_\varepsilon^{\gamma \to \alpha'}$, the mechanical energy due to the applied stress. Therefore M_s in the unstrained system can be moved to a higher temperature, M'_s, where the sum of the mechanical and thermodynamic energies $(\Delta G_\varepsilon^{\gamma \to \alpha'} + \Delta F^{\gamma \to \alpha'}/M'_s)$ is equal to $\Delta F^{\gamma \to \alpha'}/M_s$, the critical undercooling energy at M_s. Thus cold working at

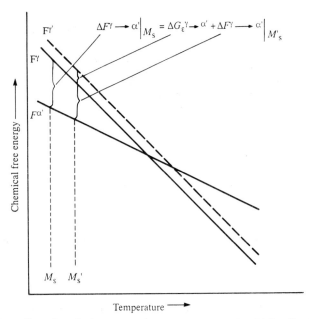

Figure 5.25   *Effect of applied stress on the martensite reaction (After Patel and Cohen[27])*

temperatures above M_s may result in the formation of *strain-induced martensite*, depending upon the chemical composition of the steel and the temperature of working.

In addition to their influence on the martensite reaction, alloying elements can also affect the work-hardening characteristics of austenitic stainless steels through their effect on the *stacking fault energy* of the system. Stacking faults are planar imperfections in the normal stacking of the fcc lattice which increase the rate of work hardening by hindering the movement of dislocations. Elements such as carbon, manganese and cobalt facilitate the formation of stacking faults whereas nickel and copper raise the stacking fault energy which inhibits their formation. The effects of these elements on the true stress–true strain behaviour of 18% Cr 13% Ni-base steels is shown in Figure 5.26. This base composition has sufficient stability to withstand the formation of strain-induced martensite at ambient temperature and therefore any changes that take place in the work-hardening behaviour can be attributed to stacking fault effects. Thus cobalt and manganese raise the rate of work hardening whereas nickel and copper decrease the rate. The effect of chromium on the stacking fault energy was examined in steels containing 18% Ni and, as illustrated in Figure 5.27, it would appear that chromium reduces the stacking fault energy and thereby increases the rate of work hardening.

Work hardening of commercial grades

The effects of rolling at ambient temperature on the properties of commercial grades of austenitic stainless steel are shown in Figures 5.28 and 5.29. Type 301

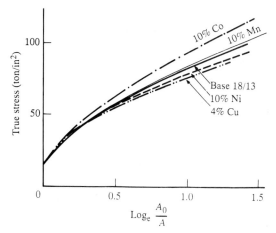

Figure 5.26   *True stress–true strain in 18% Cr 13% Ni steel with alloy additions (After Llewellyn and Murray[28])*

(17% Cr, 7% Ni) shows the highest rate of work hardening and this can be related to the fact that it has the lowest alloy content in the range examined. Therefore it has the lowest stability and forms the greatest amount of strain-induced martensite during cold rolling. Conversely, Type 316 (18% Cr, 12% Ni, 2.5% Mo) and Type 310 (25% Cr, 20% Ni) are the most highly alloyed steels in the series and these show the lowest rates of work hardening because their alloy content is sufficient to suppress the formation of strain-induced martensite at ambient temperature. However, if rolling is carried out at a lower temperature, e.g. $-78°C$, then transformation to martensite can be induced in Type 316 steel and this results in a higher rate of work hardening than that observed at room temperature. This is illustrated in Figure 5.30.

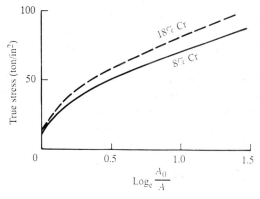

Figure 5.27   *True stress–true strain in 18% Ni base steels with different chromium levels (After Llewellyn and Murray[28])*

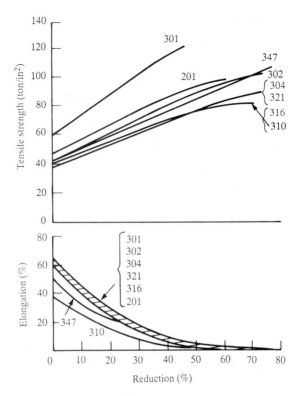

Figure 5.28 *Effect of cold rolling on the tensile strength and elongation of austenitic stainless steels (After Llewellyn and Murray[28])*

Optimization of cold-forming properties

For severe cold-forming operations, such as the cold heading of bolts, very low work-hardening rates are required. In such cases, the stainless steels selected are those which are stable enough to withstand the formation of martensite, e.g. 18% Cr 13% Ni. However, as illustrated earlier, a further reduction in the work-hardening rate can be obtained by increasing the stacking fault energy and stainless steels containing up to 4% Cu have been used to advantage in severe cold-heading operations.

   In contrast, the cold forming of stainless steel strip may require material which can undergo severe stretching. In this case, relatively high rates of work hardening are required and these are achieved via the formation of controlled amounts of strain-induced martensite. A good example is in the production of stainless steel sinks which undergo severe deformation in achieving their final shape, without the need for inter-stage annealing. As illustrated by Gladman *et al.*,[29] major changes in the rate of work hardening and uniform elongation during tensile testing can be induced by relatively minor changes in the chromium and nickel content of Type 302 (18% Cr, 8% Ni) steel. Thus the

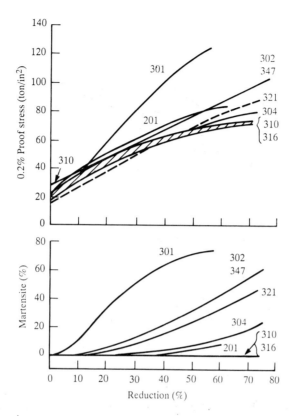

Figure 5.29 *Effect of cold rolling on the 0.2% proof stress and martensite content of austenitic stainless steels (After Llewellyn and Murray*[28]*)*

composition of stainless steel grades for sink manufacture is controlled very carefully so as to produce an optimum rate of work hardening.

Mechanical properties at elevated and sub-zero temperatures

Although austenitic stainless steels are used primarily because of their high corrosion resistance, they also possess extremely good mechanical properties over a wide temperature range. Unlike ferritic materials, austenitic stainless steels do not exhibit a ductile–brittle transition and maintain a high level of toughness at liquid gas temperatures. On the other hand, they also exhibit good creep rupture strength at temperatures above 600°C, where ferritic and martensitic steels undergo microstructural degradation. The standard grades of austenitic stainless steel are therefore used in cryogenic applications, such as liquid gas storage vessels and missiles, and at elevated temperatures in chemical and power plant applications. These aspects are discussed on pp. 264–274.

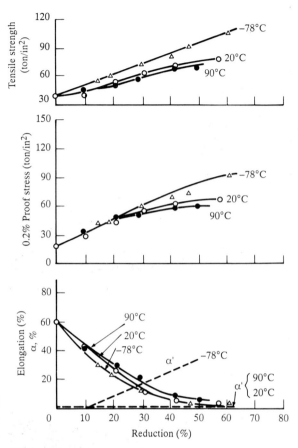

Figure 5.30   *Effect of the temperature of rolling on the properties and martensite content of Type 316 steel (After Llewellyn and Murray*[28]*)*

Tensile properties

The tensile properties of the standard grades over the temperature range -200 to $800°C$ have been examined by Sanderson and Llewellyn[30] and the data are summarized in Figure 5.31. This indicates that the tensile strengths of the majority of the grades fall within a fairly narrow scatter band at temperatures above about $100°C$. As the temperature falls below this level, there is a marked increase in tensile strength, the lower alloy steels such as Types 302 and 304 achieving higher strengths than the highly alloyed grades such as Types 309 and 310. However, this situation tends to be reversed in the case of the 0.2% proof stress values, the more highly alloyed grades such as Types 316, 309 and 310 providing higher strengths than Types 302, 304, 321 and 347. This behaviour is related to the stability of the austenitic structure in these materials and the relative ease with which they undergo transformation to strain-induced martensite.

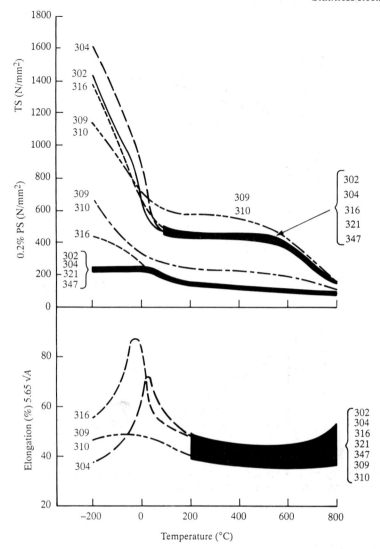

Figure 5.31   *Tensile properties of standard austenitic stainless steels: −200 to 800°C (After Sanderson and Llewellyn[30])*

At temperatures above 100°C, solid solution strengthening is the main criterion controlling strength and therefore the more highly alloyed steels, such as Types 309 (25% Cr, 15% Ni) and 310 (25% Cr, 20% Ni), possess the highest strengths. However, at temperatures below 100°C, deformation in the tensile test takes place by means of:

1. Slip/dislocation movement.
2. Transformation to strain-induced martensite.

Thus the more lowly alloyed grades such as Type 304 (18% Cr, 9% Ni) undergo strain-induced transformation to martensite with relatively little strain at temperatures below about 50°C, resulting in low proof stress values. However, this behaviour also leads to a very high rate of work hardening and produces high levels of tensile strength. Conversely, steels such as Types 309 and 310 are stable enough to resist the formation of strain-induced martensite and their proof stress values increase significantly below 100°C, following the characteristic of fcc alloys in showing a marked increase in flow stress with decreasing temperature. However, the suppression of strain-induced martensite leads to low rates of work hardening and therefore to relatively low levels of tensile strength.

As indicated in Figure 5.31, the elongation values of these materials are also strongly correlated with alloy content at temperatures below about 100°C. Thus stable grades such as Types 309 and 310 exhibit a modest peak in ductility at temperatures of the order of −50°C whereas the less stable grades such as Types 304 and 316 show sharply defined maximum values at temperatures of about +20 and −20°C respectively. The reason for this behaviour is complex but can be explained in terms of the temperature dependence of the flow stress and work hardening rate.[30]

Impact properties

The impact properties of the standard austenitic grades over a similar range of temperature are shown in Figure 5.32 (a) and (b). Again with the exception of Types 309 and 310, the materials fall within reasonably well defined scatter bands but with a very clear differentiation between Charpy and Izod values. In the Charpy test, the energy values decrease at temperatures below about 100°C, whereas in the Izod test the values decrease progressively as the temperature is increased. However, these variations are largely a function of the operating conditions in the testing machines used for this work, and the main point to emerge is the excellent toughness provided by these steels at extremely low temperatures.

Steels for boilers and pressure vessels

Carbon, alloy and stainless steels are used in boilers operating at elevated temperatures and also in pressure and containment vessels operating at room and sub-zero temperatures. Whereas the metallurgy and applications of these various types of steel have been discussed in discrete chapters of this text, it was considered more appropriate to deal with this general topic under a single heading at this stage. Thus, although this chapter is concerned primarily with stainless steels, this particular section will also deal with the use of carbon and alloy steels in boilers and pressure vessels.

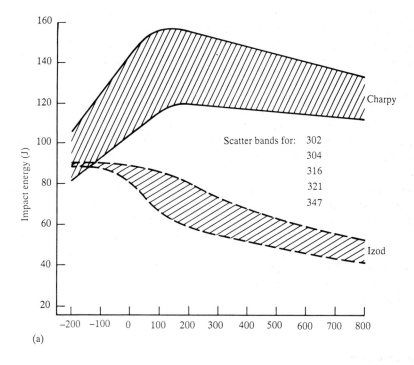

(a)

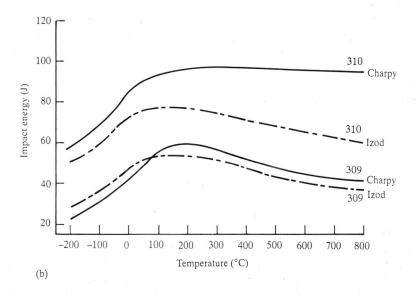

(b)

Figure 5.32 *Impact properties of standard austenitic stainless steels: − 200 to 800°C (After Llewellyn and Sanderson[30])*

Steel specifications

Pressure vessels are manufactured from plates, sections, tubes and forgings and steel specifications have been formulated specifically for these and other product forms for use in pressure vessels. A summary of the relevant BS specifications in this sector is given in Table 5.12. Thus, whereas plates for bridges and buildings are specified in BS 4360 (Weldable structural steels), similar grades of steel in the form of plates are also identified in BS 1501 (Steels for fired and unfired pressure vessels). The first reaction to this situation might well be to suggest that this represents an unnecessary and confusing duplication but in fact the two specifications deal with the property data that are important to the intended applications. Whereas both specifications cover tensile properties and the impact strength at room and sub-zero temperature, BS 1501 also specifies tensile and stress rupture properties at elevated temperatures.

Pressure vessel codes

In an earlier chapter, an introduction was given to the British Standard Specification relating to the construction of bridges (BS 5400) and steelwork in

Table 5.12   *BS Specifications for boiler and pressure vessel steels*

BS	Product Form	Title
1501	Plates	Steels for fired and unfired pressure vessels
		Part 1: C and C–Mn steels
		Part 2: Alloy steels
		Part 3: Corrosion and heat resisting steels
1502	Sections and bars	Steels for fired and unfired pressure vessels
1503	Forgings	Steel forgings for pressure purposes
1504	Castings	Steel castings for pressure purposes
3059	Tubes	Steel boiler and superheat tubes
		Part 1: Low tensile carbon steel tubes without specified elevated temperature properties
		Part 2: Carbon, alloy and austenitic stainless tubes with specified elevated temperature properties
3601	Pipes and tubes	Steel pipes and tubes for pressure purposes: carbon steel with specified room temperature properties
3602	Pipes and tubes	Steel pipes and tubes for pressure purposes: carbon and carbon manganese steels with specified elevated temperature properties
3603	Pipes and tubes	Steel pipes and tubes for pressure purposes: carbon and low alloy steel with specified low temperature properties
3604	Pipes and tubes	Steel pipes and tubes for pressure purposes: ferritic alloy steel with specified elevated temperature properties
3605	Pipes and tubes	Seamless and welded austenitic steel pipes and tubes for pressure purposes
3606	Tubes	Steel tubes for heat exchangers

buildings (BS 5950). In a similar vein, specifications have also been prepared for the construction of boilers and pressure vessels, e.g.:

1. BS 113: 1992 Design and manufacture of water tube steam generation plant.
2. BS 5500: 1994 Unfired fusion welded pressure vessels.

Both specifications have a materials section which defines the steels that can be used and the associated design stress values. An example is shown in Table 5.13 which has been reproduced from BS 1113. The steels identified in this table are from BS 3059 (Steel boiler and superheat tubes) and cover a range of compositions from 0.3% Mo up to 12% Cr–Mo–V. Against each steel grade, a series of design stress values (f) is provided for various operating temperatures up to a maximum of 620°C in this particular table. As illustrated shortly, these design stresses are calculated on the basis of short-term tensile properties at low to moderate temperatures, but at higher operating temperatures design is based on creep rupture strength. Where the latter is the case, the values in Table 5.13 are shown in italics.

Thus specifications such as BS 1113 and BS 5500 provide design engineers with clearly defined data on the allowable design stress values at different operating temperatures for the various steel specifications and product forms identified in the previous section.

Steels for elevated-temperature applications

Prior to dealing with the derivation of design stress values, the following terminology must be defined:

- R_m min. TS at room temperature
- $R_{e(T)}$ min. 0.2% PS (1.0% PS for austenitic steels) at temperature T
- S_{Rt} mean value of stress to cause rupture in time t at temperature T
- f_E design strength based on short-term tensile strength properties
- f_F design strength based on creep rupture characteristics

In BS 1113 and BS 5500, the allowable design stresses at elevated temperatures are calculated in the following manner:

1. C, C–Mn and low-alloy steels

BS 1113

$$> 250°C \; f_E = \frac{R_{e(T)}}{1.5} \text{ or } \frac{R_m}{2.7}$$

BS 5500

$$\geqslant 150°C \; f_E = \frac{R_{e(T)}}{1.6} \text{ or } \frac{R_m}{2.35}$$

whichever gives the lower value.

Table 5.13 Design stress values for tube steels according to BS 1113 (*Design and manufacture of water tube steam generating plant*)

Type–grade and method of manufacture	R_m (N/mm²)	R_e (N/mm²)	Thickness (mm)	250	300	350	400	440	450	460	470	480	490	500	510	520	530	540	550	560	570	580	590	600	610	620	Design lifetime (h)
BS 3059: Part 2 – Alloy steel																											
243 0.3Mo S1, S2 ERW and CEW	480	275		149	128	120	117		115			113	112	94	78	62	52	41	32								100000
													107	87	70	57	46	37	30								150000
													100	81	65	53	42	34	28								200000
													95	77	62	50	40	32	25								250000
620 1Cr½Mo S1, S2 ERW and CEW	460	180		129	129	129	120		116			114	113	112	93	76	62	52	42	33	27						100000
													113	102	83	67	55	44	35	29	25						150000
													113	94	76	61	49	40	32	26	22						200000
													107	88	70	57	45	37	30	25	20						250000
622–490 2¼Cr1Mo S1, S2	490	275		157	153	149	145		137	135	133	131	118	105	94	82	72	61	53	45	39	34	29	26			100000
											133	122	108	97	85	73	63	56	48	42	36	31	27	23			150000
											130	117	104	92	79	68	59	52	45	38	33	28	25	22			200000
											126	113	100	87	75	65	57	49	42	36	32	27	23	20			250000
629–590 9Cr1Mo S1, S2	590	400		219	217	215	211	208	207	207	195	176	159	144	129	115	103	91	80	68	58	48	38	31	26		100000
										207	188	169	152	137	123	110	97	85	74	62	52	42	34	28	23		150000
										204	183	165	148	132	118	105	93	81	69	58	48	38	31	25	22		200000
										199	179	159	144	129	115	102	89	78	66	55	45	35	28	24	21		250000
762 12CrMoV S1, S2	720	470		241	234	230	225		215	205			201	191	173	155	138	122	107	93	80	68	58	48	40	33	100000
													200	184	168	152	135	115	98	85	72	62	52	44	37	30	150000
													197	180	164	146	128	110	94	80	68	58	49	41	34	28	200000
													192	176	160	142	124	105	90	77	65	55	46	38	32	25	250000

After BS 1113: 1989.

2. Austenitic stainless steels

BS 1113

$$> 250°C \, f_E = \frac{R_{e(T)}}{1.35} \text{ or } \frac{R_m}{2.7}$$

BS 5500

$$\geqslant 150°C \, f_E = \frac{R_{e(T)}}{1.35} \text{ or } \frac{R_m}{2.5}$$

whichever gives the lower value.

3. For time-dependent design stresses (creep range)

BS 1113 and BS 5500

$$f_F - \frac{S_{Rt}}{1.3}$$

These rules give rise to the type of design curve shown in Figure 5.33 and which is a graphical representation of the sort of information that is presented in Table 5.13. As illustrated schematically in this figure, there is a marked change in the design stress values at the *cross-over* temperature where the values based on short-term tensile properties give way to those based on the time-dependent creep rupture strength.

The actual design stress values for a series of ferritic grades and austenitic stainless steels are shown in Figure 5.34(a) and (b) respectively. The steels concerned are in tube form (to BS 3059) and the design values in the creep range relate to rupture in 100 000 hours. As illustrated in Figure 5.34(a), the allowable stresses increase progressively as the alloy content of the ferritic grades is

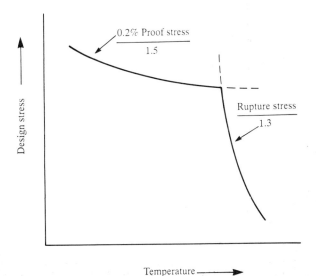

Figure 5.33 *Design strength values for low-alloy steels according to BS 1113*

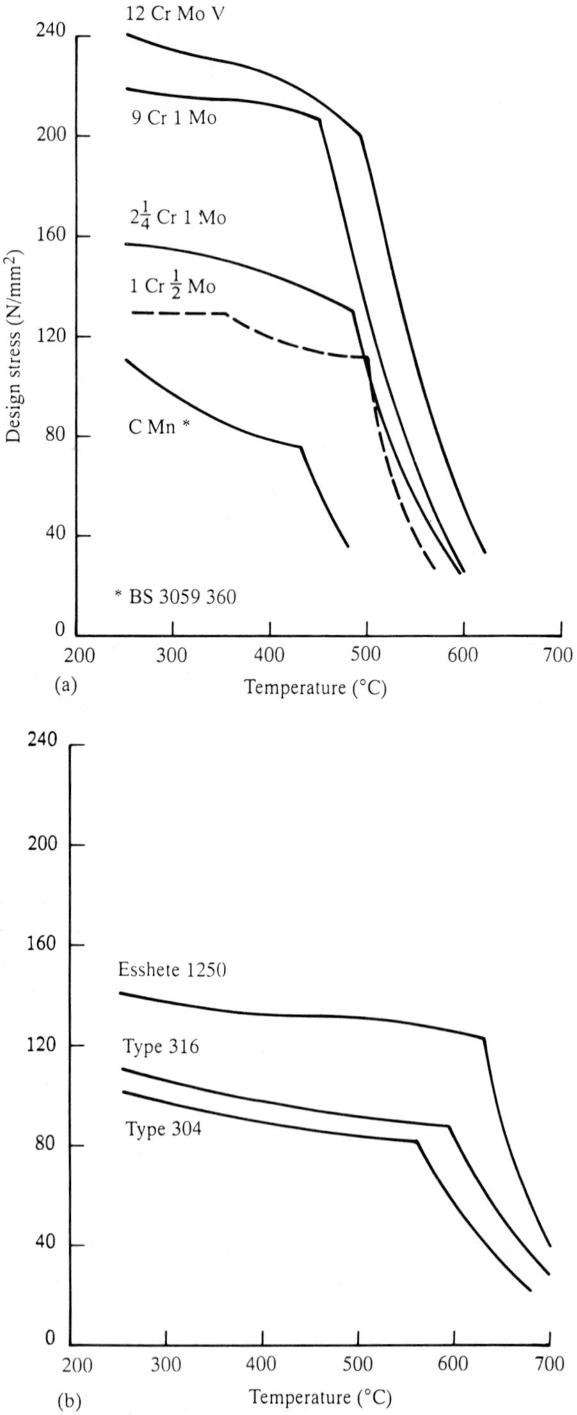

Figure 5.34   *Design stress values according to BS 1113: 1992 (Design and manufacture of water tube steam generating plant); (a) C–Mn and alloy steels; (b) austenitic stainless steels*

increased. Whereas the C–Mn steel exhibits a cross-over temperature of 430°C, the alloy steels undergo this change in the range 450–500°C. In the austenitic series, Figure 5.34(b), the design stress values increase progressively in the order:

- Type 304 – 18% Cr, 9% Ni
- Type 316 – 18% Cr, 12% Ni, 2.75% Mo
- Esshete 1250 – 16% Cr, 10% Ni, 8% Mn–Mo–V–Nb

The last composition was developed specifically for use in power plant boilers and will be discussed later on pp. 278–280. However, the main feature to note with these steels is that the design stress values are relatively low at temperatures up to about 500°C compared with the ferritic grades, whereas the cross-over between short-term tensile properties and time-dependent properties is extended to considerably higher temperatures. As discussed later, Esshete ·1250 offers a significant advantage over standard grades because of the facility to design on the basis of proof strength at operating temperatures up to 630°C.

From the data presented in Figure 5.34, it can be appreciated that the economic selection of steels for service at elevated temperature would dictate the use of C–Mn or low-alloy grades at temperatures up to 500–525°C. However, at higher temperatures, use must be made of austenitic stainless grades which exhibit better properties, particularly in the creep rupture range.

Steels for low-temperature applications

As illustrated on pp. 264–265, austenitic stainless steels possess excellent impact properties at extremely low temperatures, which is a characteristic of face-centred-cubic metals. Therefore stainless steels, together with alloys of aluminium, copper and nickel, are used extensively for the construction of equipment operating at temperatures below −120°C. However, there are many structures which operate at only moderately low temperatures which do not require the high toughness levels of these materials and which are satisfied by the use of less expensive steels. This situation is demonstrated very clearly by Wigley[31] in Figure 5.35, which identifies the boiling temperatures for a wide range of gases and the materials that are used for the associated structures and containment vessels. Thus for temperatures down to about −50°C, C–Mn steels (to BS 4360) may be adequate and a range of nickel steels ($2\frac{1}{4}$, $3\frac{1}{2}$ and 9% Ni) can be used progressively down to temperatures of the order of −200°C. Wigley states that 9% Ni steel is the only ferritic steel which is permitted for use at liquid nitrogen temperatures (−196°C). It is also economical for the construction of storage tanks for liquid argon (−186°C), oxygen (−183°C) and methane (−161°C).

The specifications, composition ranges and properties of these C–Mn and nickel steels are summarized in Table 5.14. In this table, Wigley makes the point that although specific impact test temperatures are laid down in the relevant standards, this does not guarantee that the steel will be satisfactory for use at

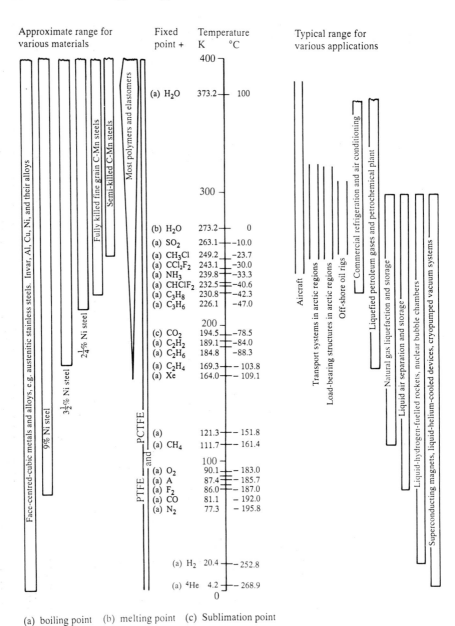

(a) boiling point   (b) melting point   (c) Sublimation point

Figure 5.35   *Some significant temperatures and typical operating temperature ranges for various structures and constructional materials (After Wigley[31])*

those particular temperatures. This aspect is particularly pertinent if the materials are used in thick sections. Appendix D of BS 5500 (1994) therefore specifies particular requirements for ferritic steels for use in vessels operating below 0°C. These take the form of diagrams providing relationships between the lowest operating temperature (design reference temperature) and impact test

Table 5.14 The compositions and mechanical properties of ferritic steels for use at low temperatures

Type	Condition	Designation		C (%)	Mn (%)	Si (%)	Mo (%)	Nb (%)	Ni (%)	Cr (%)	Yield strength (MPa)	Tensile strength (MPa)	Elongation (gauge length $5.65\sqrt{S_o}$) (%)	Charpy V-notch impact test temperature[a] (°C)
							Composition				*Mechanical properties at 20°C*			
C–Mn semi-killed	As rolled	BS 4360:	grade 40C	0.22	1.6						260	400–480	25	0
			grade 43C	0.22	1.6						280	430–510	22	0
C–Mn–Nb semi-killed	Normalized	BS 4360:	grade 40D	0.19	1.6			0.1			260	400–480	25	−20
			grade 43D	0.19	1.6			0.1			280	430–510	22	−20
			grade 50C	0.24	1.6	0–0.55		0.1			355	490–620	20	−15
C–Mn–Nb silicon killed	Normalized	BS 4360:	grade 50D	0.22	1.6	0.1–0.55		0.1			355	490–620	20	−30
C–Mn silicon killed, fine grain	Normalized	BS 4360:	grade 40E	0.19	1.6	0.1–0.55					260	400–480	25	−50
			grade 43E	0.19	1.6	0.1–0.55					280	430–510	22	−50
			grade 55E	0.26	1.7	0–0.65		0.1			450	550–700	19	−50
Ni–Cr–Mo	Quenched and tempered	BSC QT 445 ASTM A517 Gd F (USS T1)		0.1–0.2	0.6–1	0.15–0.35	0.4–0.6		0.7–1	0.4–0.65	690	795–930	18	−45
2¼%Ni	Normalized	ASTM A203 Gd A		0.17–0.23	0.7–0.8	0.15–0.3			2.1–2.5		255	450	25	−60
3½%Ni	Normalized and tempered	BS 1501/503 ASTM A203 Gd D		0.17–0.2	0.7–0.8	0.15–0.3			3.25–3.75		255	450	22	−100
9%Ni	Double normalized and tempered	BS 1501/509		0.13	0.9	0.13–0.32			8.4–9.6		515	690	22	−196
	As welded with Ni–Cr–Fe (Inconel 92) electrodes	ASTM Code Case 1308 and A353									480	655	–	−196

[a] The impact test temperature is that laid down by the relevant standards for quality control purposes. This does not guarantee that the steel, especially if used in thick sections, will be completely satisfactory for use at low temperatures. (For further details see text.)

After Wigley.[31]

temperature for different thicknesses of material. These relationships are shown in Figures 5.36(a) and (b) for as-welded and post-weld heat-treated conditions respectively. Given the beneficial effect of post-weld stress-relieving treatments, the impact test requirements are less demanding than those required in the as-welded condition. For example, 20 mm material operating at a minimum design temperature of $-40°C$ would need to provide a minimum Charpy V value at a test temperature of $-50°C$ in the as-welded condition compared with a temperature of $0°C$ in the post-weld heat-treated condition. The minimum Charpy V value depends upon the strength of steel employed as indicated in Table 5.15.

Table 5.15

Specified min. tensile strength (N/mm^2)	Min. impact energy (J) at material test temperature			
	10 mm × 10 mm	10 mm × 7.5 mm	10 mm × 5 mm	10 mm × 2.5 mm
<450	27	22	19	10
≥450	40	32	28	15

Steels in fossil-fired power plants

Since the early 1970s, major events have taken place which have had a profound effect on the choice of fuels for power generation. First, the oil crises of 1973 and 1979 curtailed the use of oil for steam-raising purposes but, currently, this fuel accounts for about 8% of power generation. At a later stage, the fire at Three Mile Island in the United States and the devastating *melt-down* at Chernobyl had a very significant effect in limiting further commitment to nuclear power. More recently, the privatization of the electricity industry in the UK has prompted the *dash for gas* and in 1993, natural gas accounted for 9.2% of power generation. As a result, the use of coal is declining but, currently, coal makes up 53% of the power requirements. However, with proven reserves for more than 300 years, coal-fired power stations are likely to feature prominently well into the twenty-first century.

Boiler layout and operation

The selection of steels for the various parts of fossil-fired boilers will be appreciated more readily by first providing an outline of boiler construction and the steam–water circuit. A schematic cross-section through a boiler is shown in Figure 5.37. The boiler has a large rectangular combustion chamber in which high-temperature gases are generated by burning coal or oil, providing maximum temperatures of up to 1600 and 2000°C respectively. The walls of the furnace section are formed with the *evaporator* tubes which are arranged in closely packed, vertical rows. Each tube has two longitudinal fins at 180° and

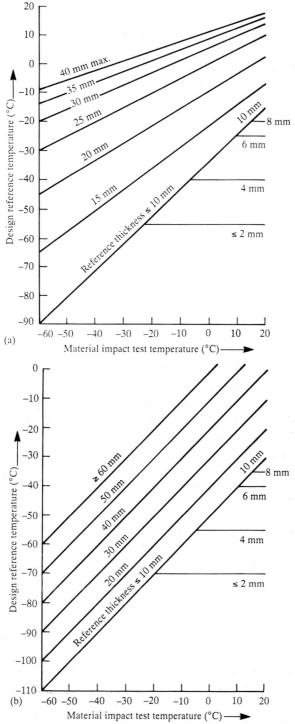

Figure 5.36 *Relationship between minimum operating temperature and impact test temperature for various thicknesses of ferritic steels: (a) as-welded condition; (b) post-weld heat-treated condition (From Appendix D, BS 5500: 1994)*

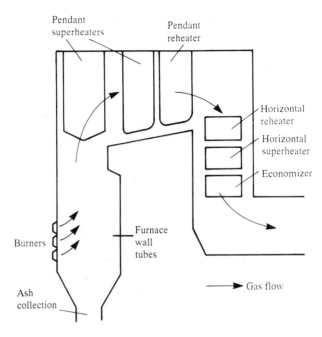

Figure 5.37   *Schematic illustration of coal-fired boiler*

the fins on adjacent tubes are welded together to form an airtight wall, except for entry ports for burners or exits for flue gas.[32] Although the temperatures of the furnace gases are very high, the cooling effect of the steam/water mixture generally limits the outer metal (*fire-side*) temperature of the evaporator tubing to about 450°C. On the other hand, significantly higher metal temperatures can be generated in some circumstances, e.g. when a flame impinges directly onto the surface of the evaporator tubes.

The atmosphere in the combustion chamber of fossil-fired boilers can be extremely aggressive, leading to severe attack and premature replacement of the boiler tubes. This arises from the chlorine content of coals whereas sulphur and vanadium are the damaging elements in fuel oils. Under normal oxidizing conditions, a protective oxide scale is formed on the surface of the tubing which limits the corrosion rate. However, reducing conditions promote complex reactions between sulphur, chlorine, carbon and oxygen, leading to the formation of non-protecting scales. Particularly damaging conditions are produced when unburnt particles of coal impinge on the side-wall of the evaporator tubes, the generation of carbon monoxide promoting sulphidation and the formation of hydrogen chloride.[33]

At the top of the combustion zone, the hot gases are turned through an angle of 90° and over banks of *superheater* tubes which are suspended from the boiler roof. In this region of the boiler, the steam is heated progressively to maximum temperatures of 540°C in the case of oil-fired boilers and 565°C in coal-fired stations. These temperature limits in UK power stations are dictated very largely by the aggressive environmental conditions described above and the need to maintain an adequate balance between operating efficiency and excessive corrosion/oxidation of the boiler tubing. In the superheater section, the first few rows of tubing are subjected to very arduous conditions due to the impingement of hot gases and fly ash.

Beyond the superheater section of the boiler, the hot gases turn through a second angle of 90° and over the *reheater* tubes. Being in the same part of the boiler, the temperatures in the superheater and reheater are similar but, due to the fact that they operate on different steam circuits, the pressure of steam in the reheater is significantly lower than that in the superheater.

Before leaving the boiler, the gases pass over a final bank of tubes that constitute the *economizer*. At this stage, the gas temperature has fallen to about 300°C and the economizer operates at a steam temperature of about 250°C.

The flow of steam/water through the boiler is shown in Figure 5.38. Feedwater is first preheated in the economizer before passing into the evaporator tubing in the furnace walls. From the various outlets in the evaporator, the steam/water mixture then passes into a large collection vessel, known as the *steam drum*. In this vessel, the steam is separated from the water, the steam being passed to the superheater and the water being directed to the feedwater circuit. On reaching a temperature of 540–565°C and a pressure of 170 bar in the superheater, the steam is conveyed from the boiler to the turbine via *headers, steam pipes* and valves. The headers are very large pressure vessels, with many stub tube connections, and act as collection units for steam from the various tubes in the superheater. A typical unit in a large power station would measure 500 mm diameter × 40 mm wall thickness × 12 m long.

As indicated in Figure 5.38, part of the steam leaving the turbine is diverted to the reheater section of the boiler. At this stage, the steam has lost a substantial amount of pressure but, after reheating, the steam is then passed again into the turbine via a header system.

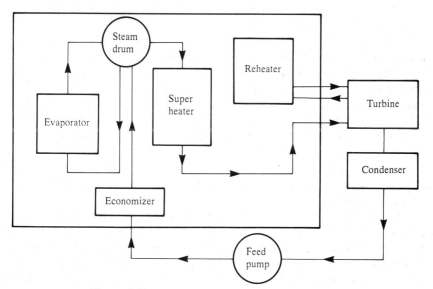

Figure 5.38 *Water–steam flow in coal-fired boiler*

The conditions described above relate to traditional coal or oil-fired stations but in recent years, the combined cycle gas turbine (CCGT) has been introduced in which the maximum steam temperature is only of the order of 400–450°C.

Steel selection

From the foregoing remarks, it can be appreciated that boiler tube materials are subjected to a wide range of temperatures and operating conditions and economic selection dictates that a variety of steel grades is used to satisfy these conditions. However, one statistic worthy of note is that in the 660 MW boilers at Drax, there are 480 km of tubing with 30 000 butt and fillet welds.

Evaporator tubing

As indicated earlier, the fire-side temperature of evaporator tubing is generally of the order of 450°C and C–Mn steel, typically BS 3059 Grade 440, is used in this region. However, in areas of the furnace wall where the operating conditions are particularly troublesome, it is sometimes necessary to use tubing which has a surface cladding of stainless steel in order to achieve reasonable life. These so-called *co-extruded* tubes will be discussed later.

Superheater/reheater tubing

Depending upon the position in the boiler and steam temperature, the tubing in various stages of the superheater will be made from the following steel compositions:

Increasing temperature resistance	1% Cr, $\frac{1}{2}$% Mo
	$2\frac{1}{4}$%, Cr, 1% Mo
	18% Cr, 12% Ni, $2\frac{1}{2}$% Mo (Type 316)
	16% Cr, 10% Ni, 8% Mn–Mo–V–Nb (Esshete 1250)

Esshete 1250 was developed by the former United Steel Companies Ltd in the 1960s specifically as a high-temperature superheater material and the number indicates the maximum operating temperature in degrees fahrenheit (675°C). As indicated in Figure 5.34, the proof strength of the steel at 600°C is 40% higher than that of Type 316, whereas its rupture strength is approximately 95% greater than that of the standard grade. This provides a design cross-over temperature from proof to rupture strength of 630°C in the case of Esshete 1250 compared with 590°C for Type 316. This enables the use of substantially thinner tubing in Esshete 1250 compared to other austenitic steels, which results in a considerable cost saving.[34]

Reheater tubing operates at the same temperature as superheater tubing and this usually dictates the use of an austenitic stainless steel, such as Type 316, in coal-fired stations. However, $2\frac{1}{4}$% Cr 1% Mo steel can be used in oil-fired boilers or in coal-fired stations where the design temperature has been reduced deliberately to allow the use of ferritic steels. In this respect, steels such as 9% Cr 1% Mo can also be considered as candidate materials.

Steam headers

Although operating at high temperatures and pressure, steam headers are not subjected to the corrosive conditions encountered by superheater or reheater tubing and therefore headers are generally made from $\frac{1}{2}$% Cr–Mo–V, 1% Cr–Mo or $2\frac{1}{4}$% Cr 1% Mo steel. For steam temperatures in excess of 565°C, highly alloyed ferritic or austenitic steels would be required, as discussed in the next section dealing with steam pipe.

Steam pipe

Steam pipes carry large volumes of steam from the headers to the turbine and, like headers, they are not subjected to aggressive environments. In the UK, $\frac{1}{2}$% Cr–Mo–V steel is used extensively for steam temperatures up to 565°C and therefore the material is operating in its creep range.

Steam pipe is heavy-walled material and, according to Wyatt,[32] $\frac{1}{2}$% Cr–Mo–V piping operating at a temperature of 565°C would have a wall thickness of 6.3 cm. At higher temperatures, greater wall thicknesses would be required or else stronger steels would be selected. These would include 9% Cr–Mo–V–Nb (T91) and 12% Cr–Mo–V (X20) which could be used in thinner sections, thereby reducing the cost of the pipework. At the super-critical Drakelow 'C' station, operating at 599°C/250 bar, both Type 316 and Esshete 1250 have been used for steam pipes.

Other components

The steam drum in the steam–water circuit between the evaporator and superheater is fabricated from C–Mn steel plates and presents no problems in material selection.

The economizer tubing, situated near the exhaust end of the boiler, operates in a gas temperature of 400°C and again a C–Mn steel is adequate for these conditions.

Feedwater pipes are made in C–Mn–Nb steel (BS 3602 Grade 490 Nb) and operate at temperatures up to 250°C. Previously, a 0.4% C–Mn steel was used for this purpose but BS 3602 Grade 490 Nb (0.23% C max.) is more weldable and has a higher proof strength than the traditional medium-carbon steel.

Co-extruded tubing

As indicated earlier, C–Mn or low-alloy steels, clad with a stainless steel or superalloys, are used in parts of the boiler where metal loss is excessive due to high-temperature corrosion. The experience of these materials in UK power stations is the subject of a detailed paper by Flatley *et al.*[35]

The tubes are manufactured by conventional hot extrusion and in essence comprise a corrosion-resistant outer layer over a conventional, load-bearing inner layer. Co-extruded tubing involving Type 310 (25% Cr, 2% Ni)/carbon steel was first introduced into the furnace wall of a 500 MW boiler in 1974 and gave a corrosion benefit of three to ten times, depending upon the severity of the location. This practice has now become widespread in UK coal-fired stations. The tube dimensions have varied from 76.2 mm OD × 8.3 mm total wall thickness (4.3 mm outer 310 on 4 mm C steel) to 50.8 mm OD × 6 mm total wall (3 mm outer on 3 mm inner).[35]

In the superheater and reheater stages of the boiler, the combination of materials has generally involved Type 310 heat-resistant steel over Esshete 1250, the high rupture strength steel. Co-extruded tubing of this type has been installed in various power stations in the UK, providing a benefit of 2.5 or more over monobloc Esshete 1250.[35] To avoid the possibility of intergranular attack due to the migration of carbon at the service temperature, later installations have incorporated niobium-stabilized, Type 310 material.

Flue gas desulphurization equipment

Prior to the privatization of the electricity industry, the UK Government announced plans for the installation of flue gas desulphurization (FGD) equipment on all new coal-fired power stations and also for the retrofitting of such plant on certain existing coal-fired stations. Following privatization, PowerGen have completed the installation of FGD at Ratcliffe. National Power have FGD operational on two units at Drax power station and plan to complete the operation by 1996. The main function of FGD plant is to remove at least 90% of the SO_2 from boiler exhaust gases and, ideally, the process should also generate a saleable by-product.

Most FGD plants operate on the basis of a wet scrubbing process in which SO_2 is absorbed in a suspension of lime (CaO) or limestone ($CaCO_3$) in water:

$$CaCO_3 + SO_2 \rightarrow CaSO_3 + CO_2$$

In the above reaction, calcium sulphite ($CaSO_3$) forms as a sludge and having no commercial value would have to be disposed of to landfill. However, by incorporating an oxidation stage in the process, the calcium sulphite is converted to gypsum:

$$CaSO_3 + O + 2H_2O \rightarrow CaSO_4.2H_2O$$

Gypsum is more environment-friendly than the calcium sulphite sludge and also represents a commercially useful product in the building industry.

The various process stages in a typical FGD are shown schematically in Figure 5.39. The flue gas leaves the boilers at a temperature of about 130°C and passes through an electrostatic precipitator for the removal of solid particles, e.g. fly ash. The gas then passes to a heat exchanger where it is cooled to about 90°C before entering a *chloride prescrubber*. UK coals contain high levels of chlorine (typically 0.25 wt %) and chloride ions are removed from the flue gas by a spray containing calcium chloride, limestone or dilute hydrochloric acid. The gas stream leaves the chloride prescrubber at a temperature of about 50°C and then enters the *SO₂ absorber* section.

As indicated earlier, the SO_2 is reacted with alkalis in this section to form gypsum. On leaving the absorber, the gas passes on to a mist eliminator for the removal of entrained droplets and then via a *reheater* to the stack.

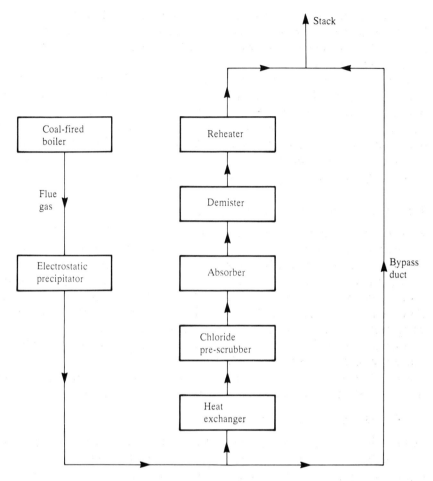

Figure 5.39 *Schematic layout for flue gas desulphurization plant*

Lane and Needham[12] have reported on the temperature and chemical conditions in the various stages of the FGD process and these are shown in Table 5.16. Prior to the heat exchanger, the gas is hot and dry and therefore the operating conditions are not aggressive. Therefore materials such as mild steel and Cor-Ten (*weathering steel*) perform adequately up to this stage.

In the heat exchanger, the conditions become more aggressive as the temperature falls below the dewpoint of the flue gas and grades such as 316L and 317LM are used in this region.

As indicated in Table 5.16, the conditions can vary significantly in the prescrubber and absorber and the choice of grade will depend very much on the particular combination of chloride ion concentration and acidity (pH). A summary of the stainless grades used in the absorber section is shown in Table 5.17. The compositions of these grades are included in Table 5.8.

Table 5.16 *Operating conditions in flue gas desulphurization plant*

Zone	Temperature (°C)	Chloride ion content (ppm)	Acidity (pH)	Comments
Electrostatic precipitator	120–150	–	–	Dry gas
Inlet duct	120–150	–	–	Dry gas
Reheater (heat extraction)	88	–	Low	Below dewpoint of flue gas plus entrained particulate matter not removed in ESP
Inlet/quench zone	88–49	High	Low	Highly abrasive, scale deposits and accumulation of chloride ions
Prescrubber	49	Up to 350000	1.5–20	Conditions will be process dependent
Absorber	49	5–30000	4.5–6.5	Conditions will be process dependent
Outlet duct	49–80	Variable	Low	Temperature below dewpoint, slurry carryover and mixing with raw gas can lead to scale deposits and high chloride levels

After Lane and Needham.[12]

Table 5.17 *Materials used in absorber towers of FGD plant*

Alloy designation	Chloride ion content (ppm)	pH	Plant
316L	1000	7.5	Cane Run 5, La Cygne 1, Cane Run 4, Colstrip 1 and 2, Duck Creek 1, Coal Creek 1 and 2, Jeffrey 1 and 2, Lawrence 4 and 5 (new) and Green River
316LM	1700	6	Dallman 3, Jim Bridger 2 and Laramie River 1 and 2
317L	Not specified		Muscatine 9
317LM	7000	6.5	Schahfer 17 and 18, Colstrip 3 and 4, Four Corners 4 and 5 and Jim Bridger 4
94L	Few thousand	6	Big Bend 4
Hastelloy G	1600	–	Cholla 1
Inconel 625	10000	2	San Juan 1, 2, 3 and 4 and Somerset 1

After Lane and Needham.[12]

Nuclear fuel reprocessing plant

The nuclear fuel reprocessing industry is a major user of stainless steels in the form of plates, tubes and forgings and the selection of steels in this sector has been reviewed very thoroughly by Shaw and Elliott.[36] Whereas the choice of materials for chemical plant is generally governed solely by the nature of the process stream, due regard must also be paid in the nuclear industry to the hazards of radiation which have a major influence on the opportunity for equipment repair and maintenance. Thus materials operating in areas of high radioactivity in nuclear reprocessing plant, where access is difficult or even impossible, are designated *primary plant materials*, whereas those operating in more accessible areas with low levels of radiation are defined as *secondary plant materials*.

Reprocessing is based upon the dissolution of irradiated nuclear fuels in hot nitric acid, followed by solvent extraction and evaporation stages. The first reprocessing plant at Sellafield in Cumbria was commissioned in 1952 and was constructed in two main types of stabilized stainless steel, namely 18% Cr 13% Ni–Nb (similar to Type 347) for primary plant and 18% Cr 9% Ni–Ti (Type 321) for secondary plant. Early work had indicated that the niobium-stabilized steel corroded at less than half the rate of Type 321 in boiling nitric acid and this was ascribed to the fact that the former had a completely austenitic structure whereas Type 321 contained some delta ferrite. However, the above authors state that it was shown subsequently that delta ferrite had no significant effect on the corrosion behaviour and that the inferior performance of Type 321 was due to the presence of TiC particles which are readily dissolved in hot nitric acid.

The use of 18% Cr 13% Ni–Nb and Type 321 remained in force until the mid-1970s, when several factors brought about a re-appraisal of steel selection. It is stated that these included the fact that some forgings in 18% Cr 13% Ni–Nb were badly corroded and that Type 321 was prone to knife-line attack in the heat-affected zone of welds. Welding problems had been experienced with the fully austenitic, niobium-stabilized steel and this material was also prone to end grain attack due to outcropping stringers of coarse NbC carbides. However, perhaps the greatest impetus for the re-appraisal of material selection stemmed from the fact that the nitric acid manufacturing industry had ceased to use the stabilized grades of stainless steels, having experienced improved manufacturing and operating performance with the low-carbon grades such as 304L and 310L.

Following extensive testing in nitric acid-based liquors and vapours, BNFL elected to replace 18% Cr 13% Ni–Nb by a low-carbon grade of 18% Cr 10% Ni steel. Initially, this material was called Nitric Acid Grade 304L but was subsequently designated by BNFL as *NAG 18/10L*. The carbon content of this grade is restricted to 0.025% max. compared with 0.03% max. in 304L and restriction on the phosphorus content (0.018% max.) also ensures very low levels of sensitization and intergranular attack. In addition, the steelmaking practice for this grade of steel has to be tightly controlled in order to produce a very clean steel so as to minimize the inclusion content and the tendency for end grain corrosion. Large quantities of *NAG 18/10L* have been used in the construction of the Thermal Oxide Reprocessing Plant (THORP) at Sellafield, and Shaw and Elliott report that the steel has given excellent welding performance. Whereas 9.3% of welds showed signs of cracking in 18% Cr 13% Ni–Nb, this has been reduced to 0.2% in *NAG 18/10L*.

Stainless steels are also used in the nuclear power and fuel reprocessing industries for the storage and transportation of spent fuel elements. For such applications, the materials must provide substantial neutron absorption, and this characteristic is provided by the inclusion of up to 1% B in a base steel containing 18% Cr and 10% Ni. The production and properties of such material, designated *Hybor 304L*, have been described by King and Wilkinson.[37] The addition of large amounts of boron to a stainless steel results in the formation of an austenite – $(FeCr)_2$ B eutectic which reduces the hot workability of the material. However, steels containing up to 1% B have been successfully rolled to plate and welded satisfactorily by the TIG, MIG and MMA processes. As illustrated in Figure 5.40, the addition of boron also produces a substantial dispersion strengthening effect, coupled with significant loss of toughness and ductility. The above authors therefore conclude that the conflicting interests of the steelmaker, fabricator and the nuclear industry are best served by steels with boron contents in the range 0.5–1%.

Corrosion/abrasion-resistant grades

Large quantities of wear-resistant materials are used for the handling of bulk solids in industries such as steelmaking, coal mining, quarrying and power generation. Many of the requirements are satisfied by quenched and tempered low-alloy steels, with hardness values in the range 360–500 BHN, and provide

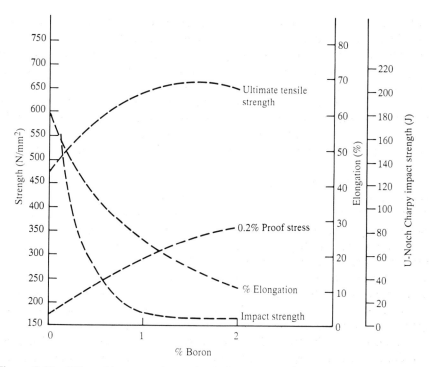

Figure 5.40 *Effect of boron on the mechanical properties of Type 304 steel (After King and Wilkenson[37])*

long life and satisfactory performance under a wide range of operating conditions. However, under wet corrosive conditions, corrosion losses can contribute significantly to the wear process. Additionally, corrosion in low-alloy steel linings can increase friction and prevent the free flow of material in chutes and bunkers, causing metering and blockage problems. Under such conditions, stainless steels become attractive alternatives to low-alloy wear-resistant grades.

One interesting example of the use of stainless steel in this field is as a composite material bonded to a thick resilient rubber backing. Both ferritic (Type 409) and austenitic (Type 304) stainless steels are used in such applications, the rubber backing reducing the amount of steel required and also the level of noise generated. Ferritic stainless steel has been used very successfully in such composites for lining coal silos in British Steel plant and many of the installations have handled over 2 million tonnes of coal.[38]

Austenitic stainless steels are generally used in situations where the conditions are very corrosive but the high rate of work hardening in these materials also provides good resistance to wear. Cook *et al.*[39] have reported on the performance of austenitic steels in a number of installations handling coal, coke and ore and, as well as providing long life compared with low-alloy steels or high-density polythene, the authors also comment favourably on the flow-promoting characteristics of Type 304 steel.

For many applications, a 12% Cr steel provides adequate corrosion resistance and the high martensitic hardness of such material also offers comparable abrasion resistance to the traditional quenched and tempered low-alloy steels. Many of the 12% Cr steels used for this type of work are based on Type 420 steel (12% Cr 0.14–0.2%C) and provide hardness levels of 400–450 BHN in section sizes up to 40 mm thick in the as-rolled condition. *Hyflow 420R* is an example of a proprietary 12% Cr steel that is marketed as a corrosion/abrasion-resisting material and provides the mechanical properties shown in Table 5.18 in the as-rolled condition. Thus low-carbon material of this type can provide a

Table 5.18

TS (N/mm^2)	0.2% PS (N/mm^2)	El (%)	2 mm CVN (J)		HV
			Long.	Trans.	
1520	1050	16	55	16	450

good combination of strength and toughness without resort to costly heat treatments.

Metcalfe *et al.*[40] have described the development of 8–12% Cr steels for corrosion/abrasion-resistant applications in South African gold mines. The rock in which the gold-bearing ore is embedded is extremely hard and causes severe abrasion problems. In addition, large quantities of water are used to allay dust and cool machinery. In turn, the water becomes contaminated with acids and salts from the working environment and, at air temperatures of about 30°C, this leads to very humid and corrosive operating conditions. The compositions of the steels developed for use in such situations are shown in Table 5.19.

Table 5.19

Code	DIN designation	Composition %					
		C	Mn	Cr	Ni	Al	N
825	X 25 CrNi 8.3	0.25		8	3		
102A	X 20 CrNiAl 10.1	0.2	1	10		0.5	
122	X 20 CrMn 12.1	0.2	1	12			
1210	X 5 CrMnN 12.10	<0.05	10	12			0.17

After oil quenching from 1100°C and tempering at 200°C, the first three steels in Table 5.19 develop microstructures that are predominantly martensitic with hardness levels in the range 471–583 HV. Steel 1210 involves an entirely different approach in that the high manganese and nitrogen contents produce an austenitic microstructure. However, the austenite is unstable and transforms to strain-induced martensite under mechanical action in service.

Automotive exhausts/catalytic converters

Motor car owners were conditioned to the fact that a traditional mild steel exhaust would need to be replaced after a typical service life of only 18 months, whereas engine and transmission components are expected to last in excess of 10 years/100 000 miles. The use of aluminized mild steel in place of uncoated material is cost-effective but, even so, the average life of the exhaust is only extended to about 30 months. On the basis of life-cycle costing, the most effective material for exhausts is a 12% Cr steel which will give a life of four to five years at a cost of about 1.5 times that of a mild steel system.

During the 1970s, British Steel directed major effort to the optimization of high-chromium steels for automotive exhausts following a major survey that showed that corrosion was responsible for about 80% of the failures that occurred in mild steel exhausts. Two types of corrosion are operative in automotive exhausts, namely:

1. That taking place on the outside of systems due to the action of water/solids thrown up from road surfaces and, more particularly, from the use of de-icing salts in winter periods.
2 That occurring on the inside of exhausts due to the action of corrosive exhaust gas condensate which forms in the cooler parts of the system.

Information was available which showed that Type 302 (18% Cr, 8% Ni) austenitic stainless steel would be very effective in overcoming these corrosion problems but it was considered that this material would prove too expensive to achieve a major market conversion from mild steel systems. Additionally, experience in the United States had indicated that a 12% Cr *muffler* grade steel might provide adequate corrosion resistance.

On the basis of preliminary laboratory tests, samples of steels containing up to 15% Cr were exposed on both the inside and outside of an exhaust system for $8\frac{1}{2}$ months, including the winter period, and during this time the car covered 16 000 miles.[41] This trial confirmed that a very significant improvement in corrosion resistance over mild steel could be achieved in steels containing 10–12% Cr and therefore further work was focused on steels within this chromium range. This culminated in the development of *Hyform 409*, a variant of the standard titanium-stabilized Type 409 grade, with the following typical composition:[42]

0.02% C, 0.6% Si, 0.3% Mn, 11.4% Cr, 0.4% Ti

Because the manufacture of exhausts involves major welding operations, a stabilized grade was required in order to eliminate any possibility of intergranular corrosion. However, as indicated earlier, the addition of titanium to a 12% Cr steel also ensures that the microstructure is ferritic, thereby providing a much more formable and weldable material than the 12% Cr martensitic grades.

Hyform 409 can be welded satisfactorily using most techniques, including

MIG, TIG and HF. When MIG welding is employed, the preferred filler metal is Type 316L. It has been established that exhausts in Hyform 409 can be manufactured on the same equipment as that used for mild steel, which is an important factor in maintaining the cost of stainless steel exhausts at a relatively low level.

Whereas 80% of failures in mild steel exhausts were attributable to corrosion, it was found that the remaining 20% were due to fatigue. Having improved the corrosion performance very substantially, fatigue therefore becomes the predominant mode of failure in stainless steels. It was established that the limiting fatigue stresses of Hyform 409 and mild steel were virtually identical. The thermal conductivity and expansion coefficients were also very similar, indicating that the 12% Cr material could replace mild steel without increasing the risk of thermal fatigue. Therefore to ensure that a stainless steel exhaust system does not undergo premature failure due to fatigue, careful consideration must be given to design details and standards of manufacture rather than to the intrinsic material properties. Extensive road trials have shown that attention to these details can provide exhausts with excellent fatigue characteristics, thereby ensuring the long-term integrity of stainless steel systems.

Whereas some car manufacturers fit complete 12% Cr exhaust systems as original equipment, others have opted for hybrid systems involving a combination of mild steel, aluminized mild steel and Hyform 409. In such systems, the 12% Cr steel is restricted to the areas that constitute the greatest risk of failure by corrosion, i.e. the rear silencer boxes.

The exhaust manifold is generally made in cast iron and the material is perfectly satisfactory in this application. However, cast iron manifolds are relatively heavy components, and for weight reduction/improved fuel economy, consideration has been given to their replacement with manifolds fabricated from stainless steel strip. Type 304 has been used for this purpose and also some of the 17% Cr steels that will be discussed in the following paragraphs on catalytic converters.

Catalytic converters have been fitted to cars in the United States since 1974 in order to reduce the level of toxic products such as CO, NO_x and unburnt hydrocarbons in the exhaust gases. These devices are also being introduced gradually in Europe and represent a major potential market for stainless steel. Typically, a platinum catalyst mounted on a ceramic substrate is contained in a stainless steel case, comprising a cylindrical shell flanked by inlet and outlet cones. In early models, the operating temperatures were low and casing materials reached a maximum temperature of 550–600°C. Given the experience with the 12% Cr *muffler* grade in the United States, steels such as Type 409 were evaluated as candidate materials for the early converters and proved to be perfectly satisfactory. However, as the requirements of legislation have tightened, the operating temperatures in the converter have been increased up to 900°C, which imposes severe demands on casing materials. These include good elevated-temperature strength to withstand the stresses produced by exhaust gas pressure and also by the weight of the device itself. Additionally, the materials are required to have good resistance to oxidation and scaling, not only for long life, but also to avoid blockage and malfunction of the converter. Type 409 has a maximum operating temperature of about 700°C and therefore more highly alloyed steels are required for modern converter systems. Whereas austenitic grades such as Type 304 (18% Cr, 9% Ni) have excellent creep rupture strength, coupled with good scaling resistance at a temperatures up to 900°C, they are expensive and their high thermal expansion characteristics could also introduce large thermal stresses and cyclic fatigue in constrained parts of the component. The lower cost and thermal expansion characteristics of ferritic steels therefore offer greater attraction as candidate materials for high-temperature converters.

The nominal compositions used in current catalytic converters are compared with Type 409 in Table 5.20.

Table 5.20

Grade	Cr%	Al%	Nb%	Ti%	Zr%
409	11.5			0.3	
430Nb	17		0.7		
18CrNb	18.6		0.6	0.3	
Armco 12SR	12	1	0.6	0.3	
Uginox FK	17		$12 \times C$		0.4

Thus the majority of the grades rely on higher chromium contents in order to achieve improved scaling resistance compared with Type 409. However, both silicon and aluminium are also beneficial in this respect and a high aluminium content is included in Armco 12SR. Each of the above steels also contains substantial amounts of strong carbide-forming elements which stabilize the materials against intergranular corrosion. However, it is interesting to note that these steels employ dual stabilization, involving niobium plus either titanium or zirconium. Whereas the latter are the more powerful stabilizing elements and also ensure a martensitic-free weld, the inclusion of niobium ensures better impact properties in the HAZ. The creep resistance of ferritic stainless steels is also improved significantly by the precipitation of Fe_2Nb (*Laves* phase). In addition to its function as a stabilizing element, zirconium is also beneficial in improving the oxidation resistance of these steels. Therefore significant benefits are obtained from the inclusion of more than one stabilizing element in ferritic stainless steels for operation at high temperatures.

Because of the very large number of cars involved and the substantial amount of material in each converter, it is reported that General Motors is now the largest consumer of stainless steel in the United States.

Architectural applications

With its aesthetic appeal, excellent resistance to atmospheric corrosion and low maintenance requirements, stainless steel offers major attractions as an architectural material. However, whereas stainless steel has featured prominently in the architectural sector in Japan and the United States, it is only recently that it has begun to enjoy similar success in the UK and Europe. One of the earliest architectural uses of stainless steel was in the cladding of the top section of the 320-m-high Chrysler building in New York in 1930. After a period of 60 years, an ASTM examination has shown that the stainless steel on this building is still in excellent condition. Of particular note in recent years have been the construction of the Lloyds building and Canary Wharf tower in London, where extensive use has been made of stainless steel cladding. In addition, stainless steel is now being used more frequently in doors, entrance halls, portals, stairways and roofing and also in street furniture such as lighting columns and telephone boxes.

As illustrated in Figure 5.41, stainless steels are now available in a range of textured finishes which increase the options open to designers and architects. Fashion and preference have also turned away from the traditional, highly reflective finishes in stainless steel in favour of the No. 4 *brushed* finish. However, this finish is produced by mechanical abrasion which can introduce fissures and crevices into the surface which can act as initiation sites for pitting corrosion. Therefore, in producing the No. 4 finish, great care has to be taken to produce a cleanly cut surface and this is generally achieved with silicon carbide belts rather than alumina abrasive belts. For architectural applications, major attention must also be given to strip flatness and to the complete removal of oxide particles from hot-band material prior to cold rolling. Small particles of scale which remain on the surface can introduce a galvanic corrosion effect which leads to rusting during atmospheric exposure.

British Steel has operated atmospheric corrosion test sites in the UK since the 1960s in various rural, industrial and marine locations. A wide range of steels and finishes are being investigated, but in the stainless steel range, attention has been focused on the following grades:

Figure 5.41   *Textured finishes in stainless steel strip (Courtesy of British Steel Stainless)*

Type 304	18% Cr	10% Ni	
Type 315	18% Cr	10% Ni	1.5% Mo
Type 316	18% Cr	12% Ni	2.5% Mo
Type 317	18% Cr	12% Ni	3.5% Mo

The stainless steel samples were examined thoroughly for evidence of pitting corrosion after 15 years exposure and the performance has been reported by Stone *et al.*[43] and also by Needham.[44] As indicated in Figure 5.42, the marine location is the most aggressive because of the high chloride content in the atmosphere. However, as anticipated, the corrosion behaviour of these steels correlates strongly with the molybdenum content. This is illustrated by the following pitting data in Tables 5.21 and 5.22 which were obtained after 15 years exposure in a marine environment.

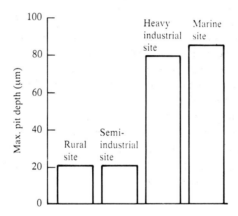

Figure 5.42   *Atmospheric corrosion in Type 304 steel after 15 years exposure (unwashed samples) (After Needham[44])*

Table 5.21

	Pit density (pits/cm²)			
	Unsheltered		*Sheltered*	
Type	*Washed*	*Unwashed*	*Washed*	*Unwashed*
304	3060	3172	3798	4214
315	647	728	789	890
316	364	377	419	450
317	37	46	52	67

After Stone *et al.*[43]

Table 5.22

Type	Maximum pit density (μm)			
	Unsheltered		Sheltered	
	Washed	Unwashed	Washed	Unwashed
304	55	85	88	102
315	37	39	50	72
316	25	24	46	70
317	22	18	25	24

After Stone et al.[43]

Thus both the maximum pit depth and the pit density decrease progressively with molybdenum content. It is also apparent that washing, either by the natural action of the rain in unsheltered conditions or by manual washing in sheltered areas, exerts a beneficial effect in reducing pitting attack. Thus the design of structures is important and, ideally, the material should be unsheltered such that rain water will remove airborne salts and other damaging debris.

The information derived from this work has confirmed that Type 316 is the most appropriate grade of stainless steel for architectural applications, particu-

Figure 5.43 *North Sea oil rig with accommodation module clad in Type 316 stainless steel (Courtesy of British Steel Stainless)*

larly where the atmosphere is relatively high in chloride ions. The long-term data obtained in marine locations were also instrumental in convincing operators of offshore oil and gas platforms of the benefits of using stainless steels for topside architecture in place of traditional painted mild steel. An example is shown in Figure 5.43. In the North Sea, the environment is particularly aggressive and the high corrosion resistance of Type 316 is attractive. However, fire resistance is a very important requirement which rules out the use of organic materials and also raises questions as to the integrity of low melting point metals such as aluminium. In the event of fire, stainless steel is also preferable to painted mild steel because of its higher strength at elevated temperature and the elimination of the smoke or toxic hazards associated with burning paint.

Because no corrosion allowance is required, stainless steel module walls can be up to 50% thinner than similar components in mild steel, resulting in significant weight savings. Whereas Type 316 stainless is considerably more expensive, the cost differential at *float out* is less than 15% compared with mild steel fabrications. A cost analysis[43] indicates that this initial differential is eliminated after the first five years of operation due to the need for the first repaint on mild steel at that time.

Based on this work, Type 316 stainless steel has now been used for wall cladding on accommodation modules, ventilation louvres and other architectural features on more than half the oil fields in the North Sea.

References

1. Irvine, K.J., Llewellyn, D.T. and Pickering, F.B. *JISI*, July, 218 (1959).
2. Andrews, K.W. *Atlas of Continuous Cooling Transformation Diagrams*, British Steel.
3. Schneider, H. *Foundry Trade Journal*, **108**, 562 (1960).
4. *Stainless Steels Specifications*, 2nd edn, British Steel Stainless.
5. Irvine, K.J., Crowe, D.J. and Pickering, F.B. *JISI*, **195**, 386 (1960).
6. Irvine, K.J. and Pickering, F.B. *ISI Special Report 86*, 34 (1964).
7. McNeely, V.J. and Llewellyn, D.T. *Sheet Metal Industries*, **49**, No. 1, 17 (1972).
8. Sedriks, A.J. *Corrosion of Stainless Steels*, John Wiley & Sons, New York (1979).
9. Fontana, M.G. and Greene, N.D. *Corrosion Engineering*, 2nd edn, McGraw-Hill (1978).
10. Copson, H.R. In *Proc. 1st Int. Cong. Met. Corr.* (London, 1962), Butterworths, p. 328 (1962).
11. Irvine, K.J., Llewellyn, D.T. and Pickering, F.B. *JISI*, October (1961).
12. Lane, K.A.G. and Needham, N.G. *Materials for Flue Gas Desulphurisation Plant No. 1 – Stainless Alloys and Coated Steels*, British Steel Stainless.
13. *Steelresearch 84–85*, British Steel, 23.
14. Binder, W.R. and Spendelow, H.O. *Trans. ASM*, **43**, 759 (1951).
15. Gregory, E. and Knoth, R.J. *Metal Progress*, January, 114 (1970).

16. *Stahl und Eisen*, **88**, 153 (1968).
17. Hooper, R.A.E., Llewellyn, D.T. and McNeely, V.J. *Sheet Metal Industries*, January, **49**, No. 1, 26 (1972).
18. Streicher, M.A. In *Proc. Stainless Steels '77*, Climax Molybdenum Co.
19. Castro, R. and de Cadenet, J.J. *Welding Metallurgy of Stainless and Heat Resisting Steels*, Cambridge University Press.
20. Gooch, T.G. In *Proc. Stainless Steels '87* (York, 1987), The Institute of Metals, p. 53.
21. Schaeffler, L.A. *Metal Progress*, **56**, 680 (1949).
22. DeLong, W.T. *Metal Progress*, **77**, 98 (1960).
23. Llewellyn, D.T., Bower, E.N. and Gladman, T. In *Proc. Stainless Steels '87* (York, 1987), The Institute of Metals, p. 62.
24. Heiple, C.R. and Roper, J.R. *Welding J.*, **60**, 74 (1981).
25. Leinonen, J.I. In *Proc. Stainless Steels '87* (York, 1987), The Institute of Metals, p. 74.
26. Kaufman, L. and Cohen, M. *Progress in Metal Physics*, Vol. 1, Pergamon Press.
27. Patel, J.R. and Cohen, M. *Acta Met.*, **1**, 531 (1953).
28. Llewellyn, D.T. and Murray, J.D. *ISI Special Report 86*, 197 (1964).
29. Gladman, T., Hammond, J. and Marsh, F.W. *Sheet Metal Industries*, **51**, 219 (1974).
30. Sanderson, G.P. and Llewellyn, D.T. *JISI*, August (1969).
31. Wigley, D.A. *Materials for Low Temperature Use*, Engineering Design Guides, Oxford University Press.
32. Wyatt, L.M. *Materials of Construction for Steam Power Plant*, Applied Science Publishers (1976).
33. Cutler, A.J.B., Flatley, A. and Hay, K.A. *The Metallurgist and Materials Technologist*, **13**, No. 2, 69 (1981).
34. Townsend, R. In *Proc. International Conference on Advances in Materials Technology for Fossil Power Plants* (Chicago, 1987) (eds Viswanathan, R. and Jaffee, R.I.), ASM International.
35. Flatley, T., Latham, E.P. and Morris, C.W. In *Proc. Advances in Materials Technology for Fossil Power Plants* (Chicago, 1987) (eds Viswanathan, R. and Jaffee, R.I.), ASM International.
36. Shaw, R.D. and Elliott, D. In *Proc. Stainless Steels '84* (Göteborg, 1984), The Institute of Metals, p. 395.
37. King, K.J. and Wilkinson, J. In *Proc. Stainless Steels '87* (Göteborg, 1984), The Institute of Metals, p. 368.
38. Guy, D.J. and Peace, J. In *Proc. Steels in Mining and Minerals Handling Applications* (London, 1990), The Institute of Metals, p. 58.
39. Cook, W.T., Peace, J. and Fletcher, J.R. In *Proc. Stainless Steels '87* (York, 1987), The Institute of Metals, p. 307.
40. Metcalfe, B., Whittaker, W.M. and Lenel, U.R. In *Proc. Stainless Steels '87* (York, 1987), The Institute of Metals, p. 300.
41. *Steelresearch 75*, British Steel, p. 43 (1975).
42. Hooper, R.A.E., Shemwell, K. and Hudson, R.M. *Automotive Engineer*, February (1985).

43. Stone, P.G., Hudson, R.M. and Johns, D.R. In *Proc. Stainless Steels '84* (Göteborg, 1984), The Institute of Metals, p. 478.
44. Needham, N.G. *Steelresearch 87–88*, British Steel, p. 49 (1988).

Index